SMP 11-16

D1742103

Teacher's guide to Book Y2

CAMBRIDGE
UNIVERSITY PRESS

The School Mathematics Project was founded in 1961 with the purpose of improving the teaching of mathematics in schools by the provision of new course materials. SMP authors are experienced teachers and each new venture is tested by schools in a draft version before publication.

Work on SMP 11–16 started in 1977 and the pilot version of the course has been used by some 50 schools, most of them comprehensive but including some selective schools, since 1980. The published version of the course started appearing in 1983.

Since its inception the SMP has always offered an 'after sales service' for teachers using its materials. If you have any comments on SMP 11–16 or would like advice on its use please write to

The School Mathematics Project,
The University of Southampton,
Southampton,
SO9 5 NH

The following people have contributed to the planning and writing of the Y, B and R series of books.

Graham Ambridge	Phil Goodwin	Brian Hughes
Chris Belsom	Eric Gower	Spencer Instone
Neil Bibby	Harry Gurevitch	Sylvia Johnson
Michael Darby	Graham Hall	John Ling
Charles Dickinson	Joyce Harris	Paul Scruton
David Fenton	Ray Harris	Martyn Truman
Tony Gardiner	Stephen Horner	Richard Walker

Others too numerous to be mentioned individually have provided valuable advice and help. Among these are the mathematics staff and pupils of the pilot schools whose detailed comments on the draft version were essential in revising the course for publication.

The SMP 11–16 team is led by John Ling.

With unfailing care the bulk of the manuscripts were typed for the press by Muriel Hudson. The authors wish to give particular thanks to Sue Glover for her work in preparing the materials for publication.

Contents

General introduction to the coloured books of SMP 11-16

The yellow, blue, red, green and amber series together make up the second part of the SMP 11-16, for pupils in years 9, 10 and 11 (ages 13+ to 16+).

The booklet strands which form the first part of the course are fully described in the *Teacher's file for key stage 3*. The relationship of the four series of books to the material in the booklet scheme is described in the *Teacher's guide to Book Y1*.

Classroom organisation

It is assumed that pupils will be grouped in sets according to ability in years 9, 10 and 11. For schools which use the booklet strands, there is guidance on setting and on the allocation of pupils to the appropriate book series in the *Teacher's file for key stage 3*. It is quite likely, whether the school uses the booklets or not, that there will be mis-matches in the setting at first, and so the setting would be regarded as provisional to start with.

Although there is rather more exposition and explanation in the books than is found in many other textbooks, the books are not intended to be 'self-instructional'. Many important points arise in the course of doing the problems in the books, and these points will need to be brought out by the teacher in discussion with the class. (This is also true of the booklets, but there the discussion is more likely to be with small groups or individuals than with the class as a whole.) Teachers may find it possible from time to time to give particular chapters or sections of chapters to the class to work through on their own, which is no bad thing since the ability to pick up information from the printed page and to follow written explanations is an important one. Where this is done it will be necessary for the teacher carefully to 'go over' what has been done. Amongst pupils working from the Y series there may be some very able ones who are quite capable of forging ahead with understanding, working from the book alone, and who should not be 'kept back'. However it is not intended that working individually through the book should be the normal method of teaching for any class, whichever series it is using. (Exceptions to this are *Books YE1* and *YE2*, and the transition books *BT* and *YT*, which are written specifically for individual use.)

There are no 'chapter summaries'. The writers feel it is more valuable for classes to make their own summary notes. The ideal ultimately is for each pupil to make his or her own notes, but initially it may be better for the teacher to lead, after each chapter, a discussion of the main ideas in the chapter, before any notes are made.

General notes for Book Y2

Mental and written arithmetic and the use of calculators

It is assumed throughout that unless there is an instruction to the contrary calculators will be used for all but the simplest calculations which can be done mentally.

We strongly recommend that teachers encourage mental calculation, and from time to time give short sets of questions to be answered mentally. We also suggest having occasional practice sessions on written arithmetic, but that the scope of these should not extend beyond addition, subtraction, multiplication by 2, 3, . . ., 9 and division by 2, 3, . . ., 9 of whole numbers and money.

Starred questions

Occasional questions are starred to indicate that they are of greater difficulty, and can be left out by slower pupils using the book.

Equipment needed for *Book Y2*

Certain standard items of equipment are needed frequently and no special attention is drawn to them in the books. These include rulers, angle measurers (recommended rather than protractors for angle measurement; see below), compasses, scissors and 2 mm graph paper.

In other cases, equipment needed (such as tracing paper) is referred to in the book. Worksheets are needed occasionally. Masters for these are available separately (see below). Three worksheets are needed for *Book Y2*, numbered Y2–1 to Y2–3. Masters for square spotty paper and triangular spotty paper, which are sometimes useful, are included in the pack of worksheet masters.

Pupils working from the Y series are assumed to have the use of a scientific calculator from *Book Y2* onwards.

Ordering equipment

The following items required for *Book Y2* are published by Cambridge University Press. You should order them through your usual school book supplier.

Worksheet masters for the Y series   ISBN 0 521 33626 0
Angle measures (pack of 5)   ISBN 0 521 24535 3

When ordering, remember to state the ISBN, the series title (SMP 11–16), the name of the item, the publisher and the number of **packs** you want. (So, for example, if you want 35 angle measurers, write your order as '7 packs of 5'.)

Notes and answers for Book Y2

1 Relationships

This chapter looks at various kinds of relationship between one variable and another. A particularly important type is the linear relationship, whose graph is a straight line. A special case of this is (direct) proportionality, where the straight-line graph goes through the point (0,0).

Proportionality is returned to in chapter 12. The more general linear relationship is considered further in a later book.

A1 The number of hours of daylight is least at the beginning of the year. It increases to a maximum after 26 weeks and then decreases again.

A2

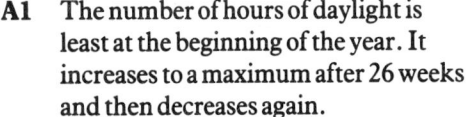

A3

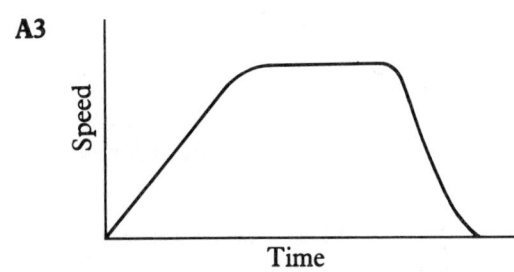

A4 (a) The students' sketches should be compared with their answers to part (d).

(b) 2·8 cm

(c)

a	0°	20°	40°	60°	80°	100°	120°	140°	160°	180°
c	0	2·8	5·5	8·0	10·3	12·3	13·9	15·0	15·8	16·0

(d)

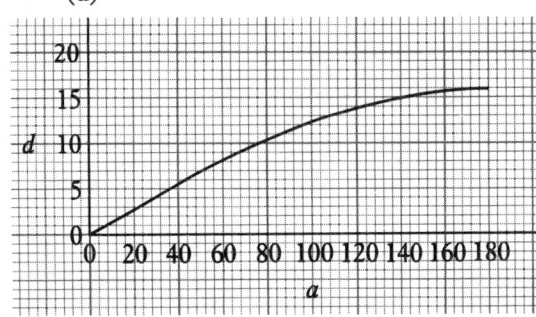

(e)

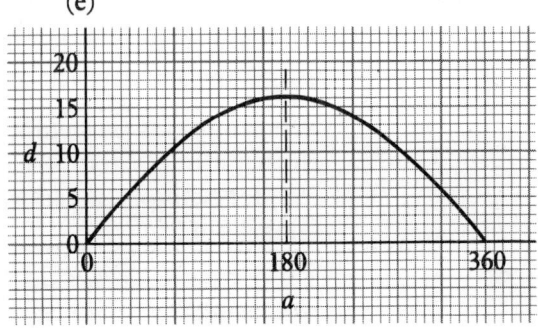

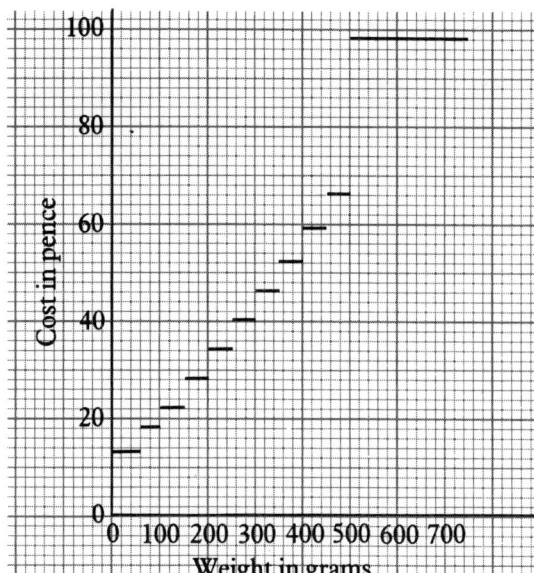

B2

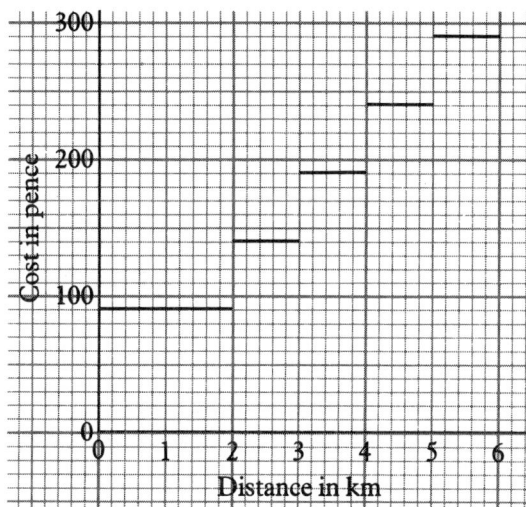

B3   (a) 2, $3\frac{1}{2}$, $4\frac{1}{2}$, 7, 8 and $9\frac{1}{2}$ minutes
  (b) It took $4\frac{1}{2}$ minutes.

B4

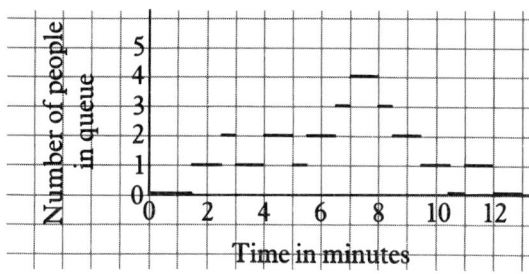

B5

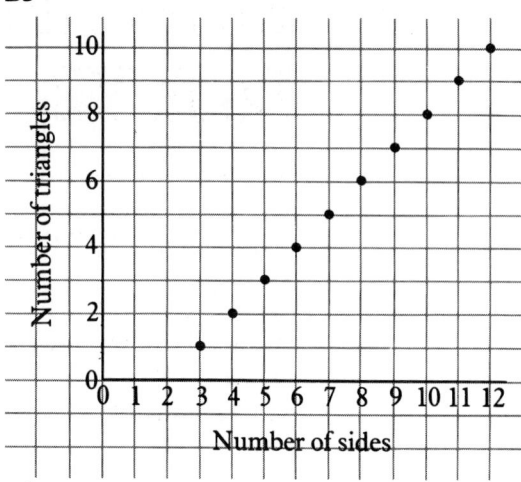

C1   (a) The length goes up by equal amounts. Yes, the relationship is linear.
  (b)

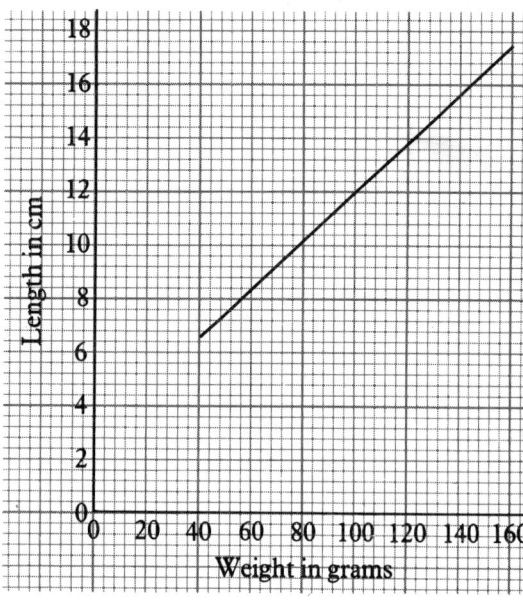

  (c) (i) About 12·5 cm
      (ii) About 85 g

C2   (a) p goes up by equal amounts but q does not.

7

(b)

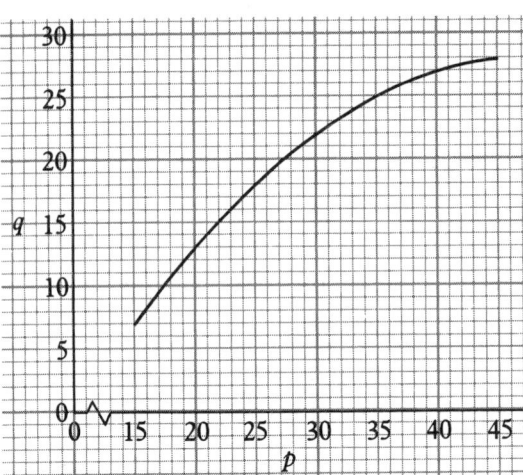

C3 (a) r goes up by equal amounts and
s goes down by equal amounts.

(b)

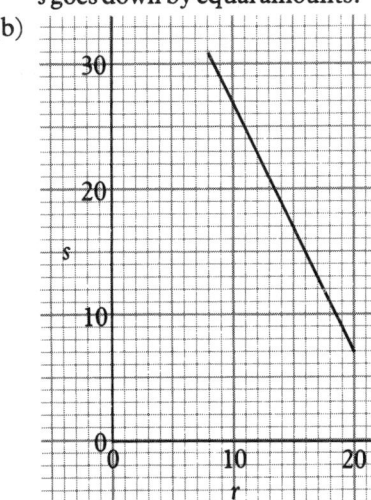

C4 (a) Yes, it is linear here, but in real life
it rarely happens.

(b)

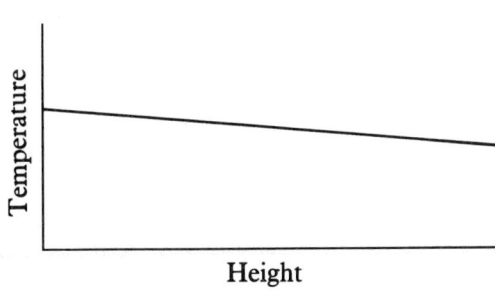

C5 (a)

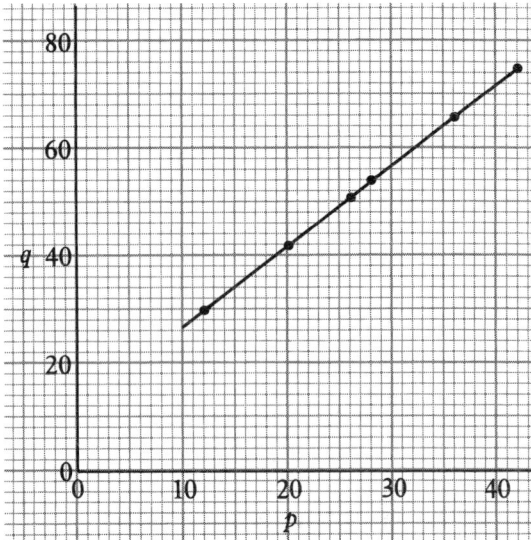

The relationship is linear.

(b)

p	10	20	30	40
q	27	42	57	72

D1 q is proportional to p in (c) and (e).

D2 (a)

Side of square, in cm	0	3	4	7	10
Perimeter of square, in cm	0	12	16	28	40

(b)

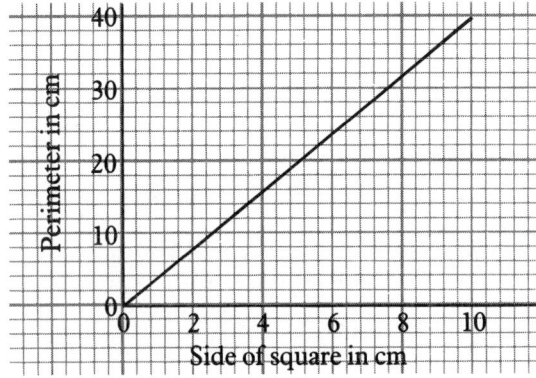

(c) Yes, it is proportional.

D3 (a)

Side of square, in cm	0	3	4	7	10
Area of square, in cm^2	0	9	16	49	100

8

(b)

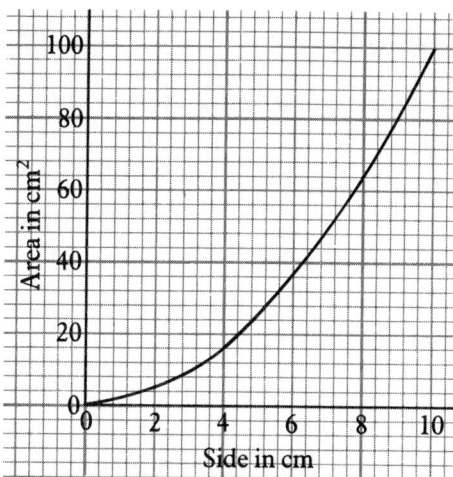

(c) No, the side of a square is not proportional to its area.

D4 Yes, the cost is proportional to the quantity.

D5 No, the cost is not proportional to the quantity bought.

D6 (a)

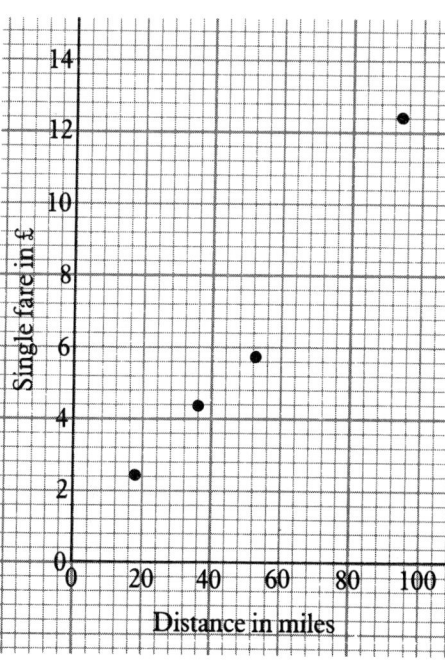

(b) No (c) Newbury (d) Westbury

D7 (a)

a	0°	15°	30°	45°	60°
h	0	13	29	50	87

(b)

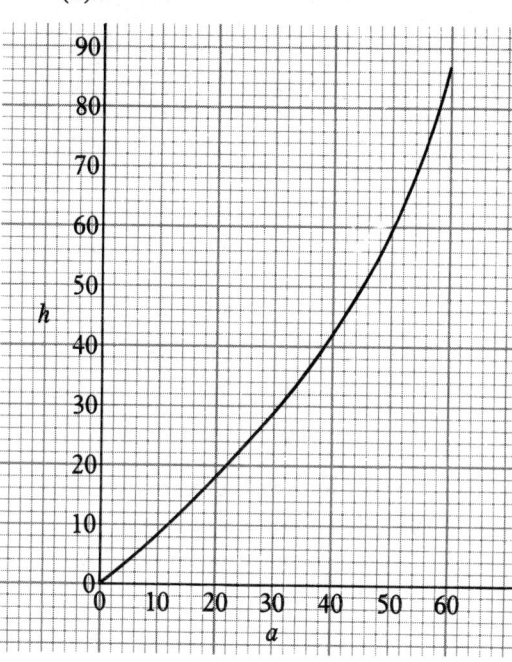

(c) No, h is not proportional to a.

D8 (a) Graph C is most likely – you have to run more slowly in long races.
(b) No, it is not proportional.

2 Accuracy

The basic idea developed in this chapter is that of an **interval approximation**: an approximation given in the form 'between … and …'. Rounded off values are equivalent to interval approximations. The later part of the chapter is about the accuracy of results obtained from calculations based on approximate data.

A1 (a) Interval length 0·1 million
    (b) Between 55·7 m and 56·3 m
    (c) Interval length 0·6 m

A2 Between 18·8 metres and 19·2 metres

A3 Between 53 000 and 55 000 increase

A4 Between 32 million and 34 million

A5 (a) Minimum 41 cm, maximum 43 cm
    (b) Minimum 5 cm, maximum 7 cm

A6 Between 6·28 cm and 6·32 cm

A7 $5·8 \pm 0·05$

A8 (a) Between 8·75 and 8·85
    (b) Between 7·635 and 7·645
    (c) Between 7·55 and 7·65
    (d) Between 2·95 and 3·05
    (e) Between 10·79 and 10·81
    (f) Between 2·795 and 2·805

A9 Between 14·90 cm and 15·10 cm

A10 Minimum 5 mm, maximum 7 mm

B1 Between 6·12 m^2 and 6·66 m^2

B2 Between 952 g and 1015 g

B3 Between 24·389 cm^3 and 29·791 cm^3

B4 Minimum 16 litres, maximum 20 litres

B5 (a) £15 (b) £300 ÷ 20
    (c) £40 (d) £400 ÷ 10

B6 (a) 0·5 (b) 0·67 (to 2 d.p.) (c) 1·5
    (d) 2·0

B7 (a) 0·67 (to 2 d.p.) (b) 1·5 (c) 0·6
    (d) 1·67 (to 2 d.p.)

C1 Between 13·55 and 13·65 grams per cm^3

C2 (a) Between 4·75 and 4·85
    (b) Between 196 500 and 197 500
    (c) Between 0·0455 and 0·0465
    (d) Between 3·795 and 3·805
    (e) Between 5·945 and 5·955
    (f) Between 3·95 and 4·05
    (g) Between 7·3095 and 7·3105
    (h) Between 2·995 and 3·005

C3

Diameter in mm	Interval approximation in ohms per m
0·02	54·85 to 54·95
0·1	2·195 to 2·205
0·2	0·5485 to 0·5495
0·5	0·08785 to 0·08795

C4 (a) City A, between 283 500 and 284 500
        City B, between 430 500 and 431 500
    (b) Between 714 000 and 716 000
    (c) No, it is not true.
    (d) No, the minimum value is 146 000 and the maximum value is 148 000.

D1 (a) 60·2008… m/s and 60·7692… m/s
    (b) 60 m/s to the nearest 1 m/s

D2 (a) Between 342·5 m and 343·5 m
        Between 212·5 m and 213·5 m
    (b) 72 781·25 m^2 and 73 337·25 m^2
    (c) 73 000 m^2 to the nearest 500 m^2

D3 540 (to 2 s.f.)

D4 (a) 5·2 (to 2 s.f.) (b) 0·065 (to 2 s.f.)

D5 340 amps (to 2 s.f.)

D6 7·9 metres (to 2 s.f.)

D7 (a) 950 (to 2 s.f.) (b) 918·75, 981·75

3 Trigonometry(1)

The approach to trigonometry is the traditional one through the ratios of sides in a right-angled triangle. The only difference is that the basic relationships are expressed in multiplicative form, e.g. $\text{adj} \times \tan \theta = \text{opp}$, rather than $\tan \theta = \dfrac{\text{opp}}{\text{adj}}$.

The multiplicative form is easier to manipulate when, for example, the length of the adjacent side is wanted.

An alternative approach is via the coordinates of points on the unit circle. In this approach there is a standard orientation, with cosine associated with measurements across, in the x-direction, and sine associated with measurements up and down, in the y-direction. Although this approach has the advantage of leading easily to sines and cosines of angles outside the range $0°$ to $90°$, it has a disadvantage in elementary applications of trigonometry where there are no standard x- and y-directions. The 'opposite, adjacent, hypotenuse' terminology has the advantage that it is carried around with the triangle, regardless of its orientation.

The relationships of cosine and sine to the unit circle and the extension to the full range of angles are dealt with in a later book.

The present chapter introduces tangent and, later, inverse tangent. We have used the notation inv tan for the latter, rather than arc tan or $\tan^{-1}$.

A1 (a) (i) AC   (ii) AB   (iii) BC
   (b) (i) DF   (ii) DE   (iii) EF
   (c) (i) GI   (ii) HI   (iii) GH

B1 (a) (i) 3·5cm   (ii) 5·1cm   (iii) 0·7
   (b) (i) 7·1cm   (ii) 10·1cm   (iii) 0·7
   (c) (i) 8·4cm   (ii) 12·1cm   (iii) 0·7
   (d) (i) 4·2cm   (ii) 6·1cm   (iii) 0·7
   (e) (i) 4·7cm   (ii) 6·6cm   (iii) 0·7

Note that the ratio calculated in **B1** and used in **B2** to **B6** appears to be accurate to 1 s.f. It is, however, accurate to 3 s.f. and answers have been given accordingly. Some students using the rule of thumb in chapter 2 may give answers to 1 s.f.

B2   2·38cm

B3   *a* 1·47m,   *b* 45·5cm,   *c* 15·4cm

B4   (a) 1·23m   (b) 0·25m   (c) 1·12m

B5   1·05m

B6   3·22m

C1   *a* 2·5m,   *b* 4·1m,   *c* 2·6m,
   d 2·6m,   *e* 0·8m,   *f* 5·2m,
   g 1·6m,   *h* 1·8m,   *i* 3·1m

D1   (a) 0·75355405   (b) 1·327044822
   (c) 0·414213562   (d) 572·9572134

D2   (a) 6·3   (b) 11·5

D3   *a* 8·2m,   *b* 12·6m,   *c* 6·2m,
   d 2·3m,   *e* 4·5m

D4   *a* 7·4,   *b* 11·1,   *c* 3·4,
   d 5·3,   *e* 8·0,   *f* 4·8

D5   *a* 9·9,   *b* 6·9,   *c* 12·9,
   d 27·9,   *e* 6·1,   *f* 5·7

D6   *a* 8·1m,   *b* 39·8m

D7   (a) 4·7m   (b) 6·3m

D8   23·3m

E1   31·0°

E2   (a) 36·9°   (b) 53·5°   (c) 4·6°
     (d) 88·5°   (e) 89·7°

E3   You may wish to discuss why a small
     discrepancy sometimes occurs.

E4 a 18·4°, b 31·0°, c 60·3°,
     d 39·4°, e 18·7°

E5 $\theta = 30°$

E6 $x = 16°$

E7 $\theta = 75·6°$

E8   (a) 51°   (b) 231°

★E9   (a) $a = 33·3°$ (b) $b = 66·5°$

★E10   72·9° and 107·1°

★E11   64·6° and 115·4°

★E12   90°, 36·9° and 53·1°

---

4 Rates

Starting with constant rates, this chapter moves on to average rates and
to values of average rates calculated from graphs.

A1   (a) 10 litre/min   (b) 67 litre/min
     (c) 12 litre/min   (d) 7 litre/min

A2   52 500 litres

A3   0·33 litre/min

A4   (a)

Time in minutes	0	0·5	1	1·5	2	3	4	5
Amount in litres	0	7	14	21	28	42	56	70

(b)

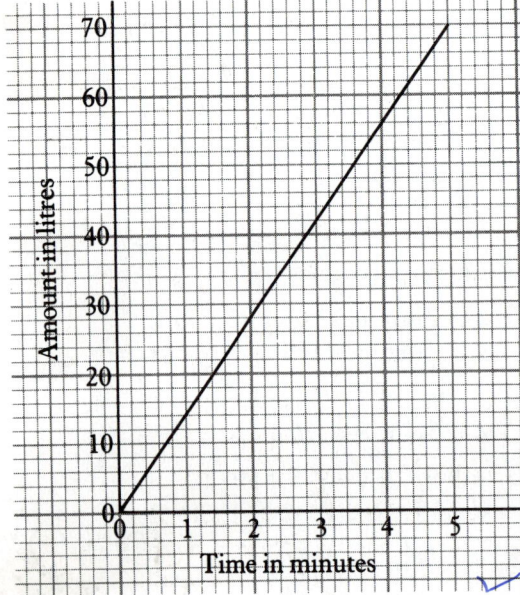

(c) 3·6 minutes

A5   62·5 litre/min

A6   (a) 15 litre/min   (b) 0·3125 litre/min
     (c) 7·5 litre/min   (d) 0·05 litre/min

A7   217 m/s

A8   6·5 m/s

A9   (a) 1·27 dollars per pound
     (b) 0·79 pounds per dollar

A10   £2·75 per hour

A11   63 pence per hour

A12   (a) 8 grams/cubic cm   (b) Density

A13   Each square metre weighs 50 grams.

B1   56·1 miles per hour

B2   47·7 miles per hour

B3   (a) 0·78 hours   (b) 6·47 hours
     (c) 2·12 hours

B4   73·5 miles per hour

B5   11·41 litre/100 km

B6   19·22 litre/100 km

B7   (a) 0·86 hundred km
     (b) 11·63 litre/100 km
     (c) 9·14 litre/100 km

C1 (a) 20 litre/min (b) 4 litre/min
(c) 10 litre/min

C2 (a) 1·6 km/min (b) 4 km/min
(c) 1 km/min

C3 (a) 250 km/h (b) 550 km/h
(c) 200 km/h

C4 1·9 km/min

C5 (a) 180 km (b) 7 hours (c) 25·7 km/h

C6 (a) 63·6 km/h (b) 47·4 km/h
(c) 44·5 km/h (d) 50·4 km/h

C7 (a) 3 degrees/min (b) 2 degrees/min
(c) 1·5 degrees/min (d) 1 degree/min
(e) 0·5 degree/min

C8 (a) 120 m (b) 12 m/s
(c) (i) 23 m/s (ii) 20 m/s (iii) 3·33 m/s
(d) 12·5 m/s

C9 (a) (i) 26·67 degrees/min
(ii) 22 degrees/min
(iii) 9·5 degrees/min
(b) (i) 24 degrees/min
(ii) 16 degrees/min
(iii) 10·67 degrees/min

D1 (a) 9 min (b) 27 litres

D2 6·5 hours

D3 4 hours

D4 (a) 58 g (b) 0·22 litre

D5 140 metres

D6 (a) 600 cubic metres
(b) In about 23 days

D7 2×10^9

D8 133·2 g of glucose per litre

D9 4·8 litres per hour

D10 1·36 minutes

5 Algebraic expressions

This chapter extends the scope of algebraic manipulation to collecting
terms and multiplying out brackets. Some of the content of this chapter
is covered in the booklets *Expressions, Brackets (1)* and
Brackets (2). Section D extends the work of chapter 9 of *Book Y1*.

A1 (a) $5a + 8b$ (b) $10p + 4q$
(c) $2x + 2y + 3z$ (d) $4h + 3k + 2l$
(e) $2s + 4t$ (f) $4m + 4n + 7p + q$

A2 (a) $11x + 7$ (b) $6p + 5q + 9$
(c) $2a + 3b + 5c + 6$ (d) $6s + t + 5$
(e) $7y + 9$ (f) $10x + 4$

A3 (a) $8x^2 + 3x$ (b) $8y^2 + 9y$

A4 (a) $7x^2 + 9x$ (b) $7pq + 2q + 7p$
(c) $a^2 + b^2 + 3ab + 2a$
(d) $4m^2 + 3n^2 + 5mn + 3m$
(e) $4f^2 + 6f$ (f) $9st + 3s + 6t$
(g) $8uv + 5u + 3v$ (h) $10a^2 + 10a$

A5 (a) $17p + q$ (b) $7b - 2a$
(c) $4x^2 + x$ (d) $4xy - 5x + 5y$

(e) $p^2 - q^2 - 5pq - 2q$ (f) $^-6a - b$
(g) $7st + 3s - 3t$ (h) $8mn - 8m - 8n$
(i) $u^2 - v^2 + 2uv - 3u - v$
(j) $3r^2 - 4s^2 + 7rs$
(k) $^-ab - 2a - b$ (l) $^-4x^2 - 4xy$
(m) $3p - q - 4pq$ (n) $5f^2 - 2e^2 - 4ef$

B1 (a) $a - b - c$ (b) $p - q - r$
(c) $x - 3 - y$ (d) $10 - s - 2t$

B2 (a) $6x - y$ (b) $3a - b$ (c) $2p - 2q$
(d) $6r - s$ (e) $3u - 3v$ (f) $3x - 2y$
(g) ^-b (h) $^-2h - 3k$ (i) $^-3p - 8q$

B3 (a) $7 - 3 + 1$ (b) $12 - 8 + 2$
(c) $10 - 6 + 2$ (d) $13 - 9 + 4$
(e) $5 - 7 + 6$ (f) $14 - 2 + 8$
(g) $10 - 8 + 2 + 1$ (h) $20 - 9 + 3 + 2$

B4 (a) $p - q + r + s$ (b) $a - b - c + d$
     (c) $e - f + g - h$ (d) $4a - 2b + 3c$
     (e) $2p - q - 3r$ (f) $5x - 2y + z$
     (g) $6p - 2q - 4r + s - t$
     (h) $5a + 2b - c - 2d - 3e$

B5 (a) $7x - 2$ (b) $2a + b$ (c) $8p - 2q$
     (d) $7s - 2t$ (e) $9u - 3$ (f) $6a + 5$
     (g) $2u + 2v$ (h) $10x + 3$

B6 (a) $3 + 5 + 2$ (b) $10 + 6 - 1$
     (c) $4 + 7 - 5$ (d) $6 + 2 - 5$
     (e) $7 + 3 + 2 + 4 + 5$
     (f) $8 + 5 - 1 + 3 - 2$

B7 (a) $p + q - r - s$ (b) $a + b + c - d$
     (c) $e + f - g + h$ (d) $4a + 2b - 3c$
     (e) $2p + q + 3r$ (f) $5x + 2y - z$
     (g) $6p - 2q + 4r - s + t$
     (h) $5a + 2b + c - 2d + 3e$

B8 (a) $2x + 2$ (b) $6a - b$ (c) $2p + 2q$
     (d) $5s + 2t$ (e) $3 - 5u$ (f) $14a - 5$
     (g) $6u + 4v$ (h) $2x + 7$

C1 (a) $2p + 2q$ (b) $2p - 2q$ (c) $3a - 6$
     (d) $20 + 4s$ (e) $3x + xy$ (f) $3ab + 2a$
     (g) $5p - 2pq$ (h) $rs - rt$

C2 (a) $4a - 2b - 6c + 10$
     (b) $6x - 6y + 12z - 6$
     (c) $10s - 4t - 4u - 4v$
     (d) $5a - 8 + 8b + 16c$
     (e) $6p - 6q + 12 - 9r$
     (f) $4a - 2b + 3ab - bc$
     (g) $5p + 3pq - pr - 2ps$
     (h) $7x - 3xy + 3y^2 - 6y$

C3 (a) $16p - 6q$ (b) $2a - 3b$ (c) $8 - 5x$
     (d) $14s - 8t$ (e) $24p - 10q$

C4 (a) $9x + 6$ (b) $15a - 4$
     (c) $4ab + 3a - 4b$ (d) $2pq - 2p - q$
     (e) $9s + 22$ (f) $5x^2 + 9x + 20$
     (g) $5x^2 + 2x + 15$ (h) $^-13x^2 + x - 12$
     (i) $p^2 - 2pq + q^2$
     (j) $3a^2 - 5ab - 4a - b - 6$
     (k) $3s^2 + 6s + 7st - 5t^2$
     (l) $x + 4y - 15$

D1 (a)

	y	$^-4$
x	xy	^-4x
3	$3y$	$^-12$

     (b) $(x + 3)(y - 4) = xy - 4x + 3y - 12$

D2 (a) $ab - 5a - 2b + 10$
     (b) $st + 4s - 5t - 20$
     (c) $pq - 6p - 4q + 24$
     (d) $xy - 7x + 3y - 21$
     (e) $fg - 2f - 2g + 4$
     (f) $m^2 - 5m - 6$

D3 (a) $x^2 + 3x - 10$ (b) $a^2 - 7a + 12$
     (c) $p^2 - 3p - 18$ (d) $2y^2 + 5y - 3$
     (e) $3q^2 - 10q - 8$ (f) $8x^2 - 18x + 9$
     (g) $9a^2 - 9a - 10$ (h) $20p^2 + 11p - 3$
     (i) $30q^2 - 49q + 20$

D4 $4x^2 - 12x + 9$

D5 (a) $4x^2 - 20x + 25$ (b) $9a^2 - 24a + 16$
     (c) $25p^2 - 20p + 4$

D6 (a) $2a^2 - 13a + 15$ (b) $9b^2 + 6b + 1$
     (c) $9c^2 - 4$ (d) $16d^2 - 40d + 25$
     (e) $6xy - 15x - 2y + 5$ (f) $9f^2 - 12f + 4$

E1 (a) $4a + 6b = 2(2a + 3b)$
     (b) $10a - 25b = 5(2a - 5b)$
     (c) $12p + 16q = 4(3p + 4q)$
     (d) $9x - 15 = 3(3x - 5)$
     (e) $35s - 28t = 7(5s - 4t)$
     (f) $18y - 6z = 6(3y - z)$

E2 (a) $3(2p + 3q)$ (b) $7(2a - 5b)$
     (c) $4(2s - 3t)$ (d) $3(f - 10)$
     (e) $2(9 - 2q)$ (f) $5(5x - 8y)$
     (g) $4(1 - 4p)$ (h) $5(4 - q)$

E3 (a) $2r + rs = r(2 + s)$
     (b) $3ab + b^2 = b(3a + b)$
     (c) $4y^2 - xy = y(4y - x)$
     (d) $2pq - p^2 = p(2q - p)$

E4 (a) $q(6p + 5)$ (b) $q(6p + 5q)$
     (c) $q(6p - 5q)$ (d) $a(2a - 9b)$
     (e) $a(5a - 3)$ (f) $b(3a + 7)$
     (g) $y(x - 3y)$ (h) $y(2x + 3z)$

E5 (a) $4b(2a + 3)$ (b) $4b(2a + 3b)$
     (c) $4b(2a - 3b)$ (d) $3a(3a - 4b)$
     (e) $5a(a - 6)$ (f) $4y(4x - 3y)$
     (g) $2a(5a - 4b)$ (h) $2ab(3a + 4b)$

Review 1

1 Relationships

1.1

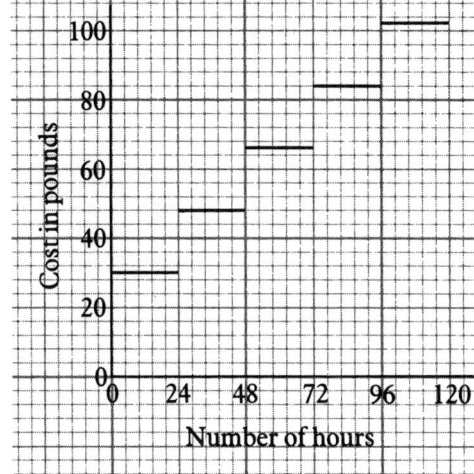

1.2 (a) No, the time for 20 swings is not proportional to the length.
The graph is not a straight line through the origin.
(b) Changing the mass on the end of the pendulum does not change the time for 20 swings.

1.3

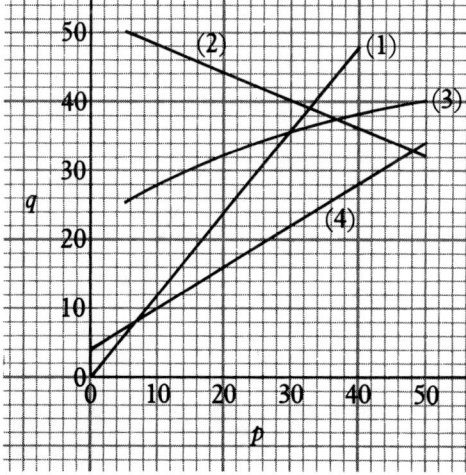

(a) In cases (1), (2) and (4)
(b) In case (1)

2 Accuracy

2.1 Between 110 and 120 km

2.2 (a) Between 5·65 and 5·75 cm
(b) Between 68·35 and 68·45 km
(c) Between 4·95 and 5·05 g
(d) Between 12 750 and 12 850
(e) Between 0·045 75 and 0·045 85
(f) Between 0·002 95 and 0·003 05

2.3 (a) 2·707 779 9 and 2·693 761 8 g/cm^3
(b) 2·7 ± 0·01 g/cm^3

2.4 9·74 cm

3 Trigonometry (1)

3.1 a 9·3 cm, b 2·1 cm, c 4·9 cm

3.2 $a = 56°$, $b = 29°$, $c = 27°$, $d = 61°$

3.3 (a) The height is 4·3 cm.
(b) The area is 10·8 cm^2.

3.4 (a) 78·8 m (b) 89·3 m

4 Rates

4.1 The super battery is better, it does 4·5 hours per penny.
The standard does 3·6 hours per penny

4.2 (a) 254 yen per dollar
(b) 0·003 94 dollars per yen

4.3 London–Glasgow, 73·4 miles per hour
London–Edingurgh, 76·1 miles per hour

4.4 (a) 10 degrees per min
(b) 1·7 degrees per min

4.5 (a) Brighton to Haywards Heath 72 m.p.h.
Haywards Heath to Gatwick 51 m.p.h.
Gatwick to Croydon 46 m.p.h.
Croydon to Victoria 32 m.p.h.
(b) 47 miles per hour

4.6 30·36 min or 30 min 21·6 sec

5 Algebraic expressions

5.1 (a) $7p - 4q$ (b) $5a - 2$
(c) $pq - 3p + 2q$ (d) $7a + 4b$
(e) $4x - 5y$ (f) $6r - 2s$
(g) $2a^2 - 7a + 2b^2$
(h) $^-pq + p - 5q + p^2$
(i) $6x^2 - 5y^2 + 6xy$
(j) $2s^2 - 3t - 4s + 5st - 7t^2$
(k) $^-5m - 3n + 5mm$ (l) $^-4a - 5 + 5b$
(m) $2x^2 + xy - 4y^2$ (n) $15p^2 - 9pq - 6q^2$
(o) $a^2 - ab + b^2 - b - a - 1$
(p) $4a^2 + 8ab + 3b^2$

5.2 (a) $4p - 3$ (b) $9y + 5$ (c) $4g + 2$
(d) $5s + 2$ (e) $15 - 4p$ (f) $u - v$
(g) $11x - 4$ (h) $11a + 3$ (i) $p - 3$
(j) $12y - 7$

5.3 (a) $3a - 6$ (b) $7x + 15$ (c) $19y - 8$
(d) $10x + 1$ (e) $13p - 34$ (f) $9a + 10$
(g) $6s - 7$ (h) $13x - 27y$ (i) $^-4b - 2a$

5.4 (a) $xy - x - 3y + 3$ (b) $a^2 + 3a - 10$
(c) $2a^2 + 9a - 5$ (d) $6p^2 + p - 12$
(e) $25x^2 - 9$ (f) $36x^2 - 12x + 1$

5.5 (a) $3(2p - 5q)$ (b) $3q(2p - 5q)$
(c) $4(1 + 2a^2)$ (d) $b(2a - 7b)$
(e) $2b(a + 5b)$ (f) $b(2a - 5c)$
(g) $5b(2a - c)$ (h) $2a(3b + 4a)$

M Miscellaneous

M1 670 m.p.h.

M2 (a) $4(4 + 2) = 24$ and $5(5 + 2) = 35$
(b) $w = 4.6$

M3 (a) 7·4 km (b) 53·3°

M4 (a) 6 routes (b) 12 routes

---

6 Trigonometry(2)

Sine and cosine are introduced in this chapter. The approach is similar to that of chapter 3.

A1 a 2·4 cm, b 3·9 cm, c 8·7 cm,
d 7·3 cm, e 8·7 cm

A2 a 7·6 cm, b 6·2 cm, c 10·3 cm
d 11·8 cm, e 10·2 cm

A3 a 8·7 cm, b 23·9 cm, c 4·4 cm
d 2·0 cm, e 3·2 cm, f 4·7 cm,
g 5·6 cm, h 4·1 cm, i 6·8 cm,
j 5·3 cm

B1 a 7·5 cm, b 6·5 cm, c 3·3 cm

B2 2·5 metres

B3 6·8 metres

C1 a 5·4 cm, b 3·3 cm, c 11·4 cm,
d 4·1 cm, e 9·2 cm

C2 A 9·8 m, B 14.0 m, C 7·8 m

C3 a 8·6 cm, b 5·6 cm, c 6·8 cm
d 6·1 cm, e 13·8 cm

C4 (a) 46 m (b) 26 m

D1 $a = 31°$, $b = 53°$, $c = 55°$
$d = 14°$, $e = 51°$, $f = 46°$

D2 67°

D3 30°

E1 37°

E2 (a) 7·0 m (b) 1·9 m

E3 (a) 8·5 m (b) 4·3 m

E4 63°

E5 (a) 69·15 m (b) 63·3 m (to 1 d.p.)

E6 5·2 cm (to 2 s.f.)

E7 (a) 23° (b) 3·9 cm (c) 7·8 cm

E8 (a) 56° (b) 9·9 cm (c) 19·8 cm

E9 12·9 cm

E10 (a) 72° (b) 11·8 cm

E11 30°

E12 39°

E13 110° and 70°

E14 59° and 121°

---

7 Investigations(1)

In investigative work the attention is focussed on processes of mathematical thinking rather than on the acquisition of specific items of content. The investigations in this chapter offer scope for discovering relationships, explaining them and using them to solve problems.

None of the material in this chapter is important as mathematical content, and no pupil's further progress will be hampered by failure to solve the problem. Thus there is no urgency to 'spill the beans' to those who have not managed it for themselves. They can be left to worry away at the problem up to the limits of their persistence. The ideas they have, even if they do not eventually lead anywhere, are often interesting and well worth discussing.

It is possible to start this work with the Koenigsberg Bridges problem (section C) and then do sections A, B and D.

A Courses for horses

A1 (a) You must either start in C and finish in E or start in E and finish in C.
    (b) A course starting in field A is not possible.

A2 Diagram 2: A and D are possible starting and finishing points.
    Diagram 3: Impossible
    Diagram 4: You can start anywhere but you must finish where you start.
    Diagram 5: A and B are possible starting and finishing points.
    Diagram 6: Impossible
    Diagram 7: Impossible
    Diagram 8: D and E are possible starting and finishing points.

A3 Diagrams 9 and 11 are impossible
    Diagrams 10 and 12 are possible.

The hint referred to in question A3 is 'Look at the number of gates round each field.' If this turns out to be insufficient, then add 'It's got something to do with odd and even.'

The rule is this: if there are more than two fields with odd numbers of gates then no route can be found. (If there are exactly two fields with odd numbers of gates then the route must start in one of these fields and finish in the other one.)

Here is the reason for the rule. If a field is 'even-gated' (has an even number of gates) the gates can be paired off. With such a field, each time we come in we can find a way out, unless we start the course in the field, in which case we have to finish in it as well. If a field is 'odd-gated', we cannot keep entering it and leaving it. Eventually we come in and have to stop, unless we start the course in the field, in which case we have to finish elsewhere. So if there are two odd-gated fields, one will be the start and the other the finish. If there are more than two, no course can be found because we cannot have more than one start and finish.

B Inspecting roads (1)

A hint, if needed, is 'Look at the number of roads which meet at each junction.'

B1   She should start at A and finish at B or vice versa.

B2   Diagram 2:   Impossible
       Diagram 3:   E and D are starting
                    or finishing points.
       Diagram 4:   E and I are starting
                    or finishing points.
       Diagram 5:   Each point could be
                    both the start and finish.
       Diagram 6:   Impossible

C The Koenigsberg Bridges problem

C1   It is not possible. There is an odd number of ways of coming or going from each piece of land and you cannot start or finish at all of them.

D Inspecting roads (2)

Although this problem can be solved 'from scratch' by trial and error, it is interesting to see whether pupils make use of what they may have learned from section B.

It is impossible for the surveyor to walk along every road once and once only because there are four junctions at which odd numbers of roads meet: A, B, C and G.

We know that at least one road will have to be covered twice. Going along a road twice (either in the same or opposite directions) is equivalent to replacing that road by two 'parallel' roads. For example,

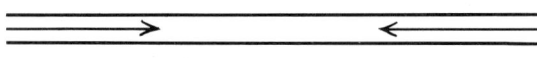

is equivalent to

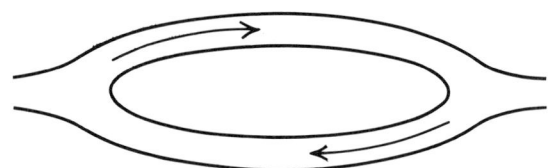

If we decide to make the 'extra road' join two of the 'odd' junctions, then the effect will be to convert these junctions into 'even' junctions, leaving only two 'odd' ones in the network.

We want to make the extra distance walked as short as possible, so the road to duplicate is the one from A to C, which is only 2 miles long.

This leaves B and G as the only remaining 'odd' junctions. So the surveyor's route must start at B and end at G (or vice versa). Its total length will be the total length of the entire network plus an extra 2 miles for covering AC twice (62 miles in all).

Even if no pupil manages to find the solution by this method, the abler ones may, after they have worked at the problem themselves, benefit from being led, with judicious hints, through the reasoning.

8 Distributions

This chapter introduces the important statistical concept of a distribution. In many circumstances the overall 'shape' of a distribution is of more fundamental interest than numerical measures, hence questions like A10 and A11. The practical work in section E is valuable and of interest in its own right.

A1

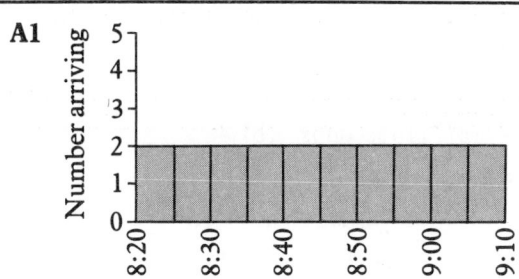

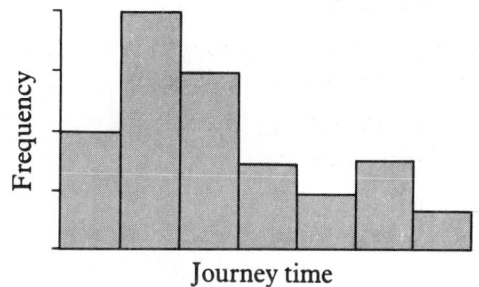

Journey time

A2 (a) 80 (b) 30 (c) 210

A3 The shortest journey time is between 0 and 5 minutes.

A4 40%

A5 14%

A6 The longest journey time is between 30 and 35 minutes

A7 The modal interval is between 20 and 25 minutes

A8 (a) 190 pupils (b) 110 pupils

A9 80% (to nearest 5%)

A10 Answers should look something like these

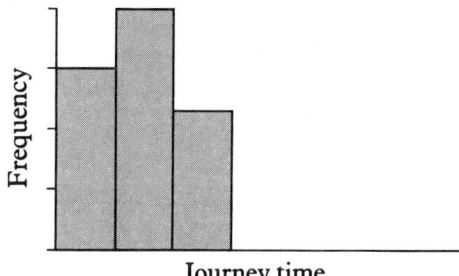

Journey time

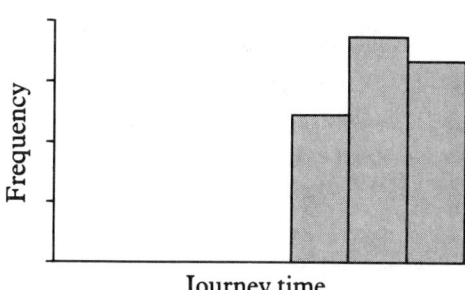

Journey time

A11 A sketch graph based on local knowledge

B1 97 kg

B2 (a) 759 kg (b) 0·5 kg

B3 17·7 °C (to 1 d.p.)

B4 (a)

Number of seeds in packet	Number of packets	Number of seeds
20	25	500
21	12	252
22	3	66
Totals	40	818

(b) Mean 20·45 seeds

B5 (a)

Weight in kg	Mid-interval value	Number of boys	Weight of group, in kg
20–25	22·5	10	225
25–30	27·5	40	1100
30–35	32·5	60	1950
35–40	37·5	50	1875
40–45	42·5	20	850
Totals		180	6000

(b) Mean 33 kg

B6 (a)

Weight in kg	Mid-interval value	Number of girls	Weight of group, in kg
20–30	27·5	20	1625
30–35	32·5	50	550
35–40	37·5	60	2250
40–45	42·5	40	1700
Totals		170	6125

(b) 36 kg   (c) Heavier, on the whole

B7 (a) 1·95 kg   (b) 412·5 kg
   (c) 2·06 kg (to 3 s.f.)

B8 (a) 116 seconds   (b) 48%

C1 Mean weight 44 g,   lightest 21 g,
   heaviest 63 g,   range 42 g

C2 (a) No, there are two distinct groups –
     adults and children
   (b) 8·95 years

D1 (a) (i) 95%   (ii) 5%
   (b) 19·2 years
   (c) The main reason is clearly that
     there are more young riders but
     it may also be true that young
     riders are more accident prone.

D2 (a) Most accidents occur in heavy
     urban traffic because motorcycles
     are used more in these conditions.
   (b) 24·9 m.p.h.

D3 (a) (i) 54%   (ii) 27%   (ii) 18%
   (b) 20%
   (c) Moderate 30%,   severe 41%
   (d) (i) 26 m.p.h.   (ii) 29 m.p.h.

E1 to E5 depend on data collected in the
     classroom.

9 Re-arranging formulas(1)

The process of re-arranging a formula is shown to be fundamentally the same as that of solving an equation, but with other variables instead of known numbers.

A1 (a) $d = 24$ (b) $d = 72·5$

A2 $t = 9·6$

A3 $s = 58·5$

A4 (a) $17 = 8 + s$ (b) $s = 9$
   (c) $38 = 5r + 3$ (d) $r = 7$

A5 (a) $20 = \dfrac{V}{4}$ (b) $V = 80$

   (c) $0·8 = \dfrac{V}{12·5}$ (d) $V = 10$

A6 (a) $19 = a - 12$ (b) $a = 31$
   (c) $17 = 62 - 3c$ (d) $c = 15$

A7 $v = 3$

B1 (a) $s = \dfrac{d}{t}$ (b) $s = 0{\cdot}35$

B2 $w = \dfrac{f}{n}$

B3 $p = q - a$

B4 $m = cn$

B5 $r = d + s$

B6 (a) $x = \dfrac{y}{3}$ (b) $x = \dfrac{y}{k}$ (c) $k = \dfrac{y}{x}$
(d) $m = g + a$ (e) $m = g - a$
(f) $a = g - m$ (g) $t = 5s$
(h) $t = sk$

B7 $m = \dfrac{p - a}{3}$

C1 $x = \dfrac{y - b}{a}$

C2 (a) $p = \dfrac{q - a}{4}$ (b) $p = \dfrac{q - a}{k}$
(c) $k = \dfrac{q - a}{p}$ (d) $a = q - kp$
(e) $f = \dfrac{m + t}{a}$ (f) $v = \dfrac{w - u}{k}$
(g) $u = w - kv$ (h) $k = \dfrac{w - u}{v}$
(i) $m = t + np$

C3 $s = p(k + a)$

C4 $c = n(d - a)$

C5 (a) $x = p(q + r)$ (b) $x = p(r - q)$
(c) $x = r(p - q)$

---

10 Points, lines and planes

The purpose of this chapter is to focus attention on some of the basic facts about points, lines and planes in three dimensions, by criticising 'impossible objects' and seeing what makes them impossible.

B1 (a) *b* and *h*; *c*, *g* and *i*; *d* and *e*
(b) *b* and *k*; *d* and *i*; *a*, *e* and *j*;
m and *r*; *c* and *l*; *n* and *q*; *h* and *g*

B2 (a), (c) and (f) show impossible objects.

C1 AF

C2 Two parallel planes never meet.

C3

(a)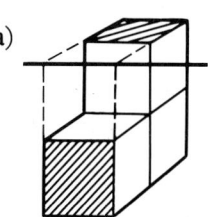

(b)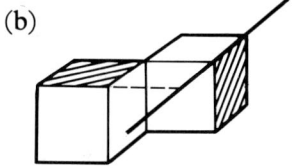

(c)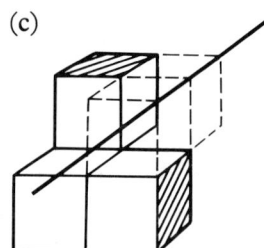

C4 (a), (c), (d) and (f) are impossible.

D1

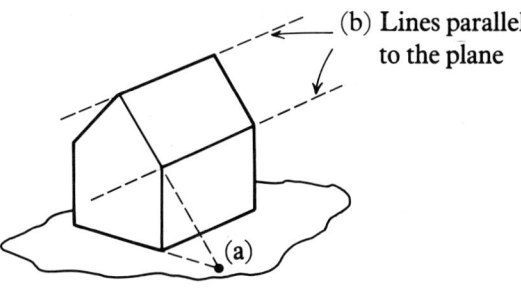

(b) Lines parallel to the plane

(a)

D2 It is not if it shows a straight line meeting a plane in two places.

D3 (a) Possible (b) Possible
(c) Not possible if they are all parts of the same line

E1 Two walls at right-angles and the ceiling, perhaps

E2 (a)

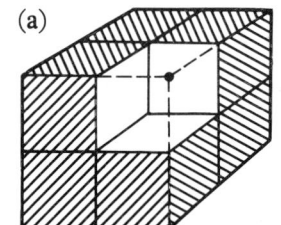

(b)

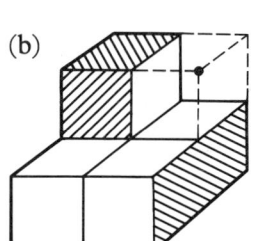

(c)

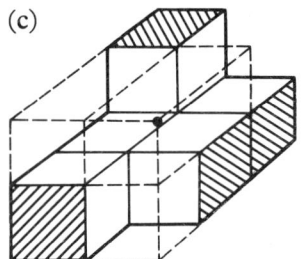

(d)

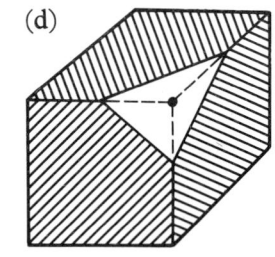

E3 (a) Possible (b) Impossible
(c) Impossible

E4 (a) Impossible (b) Impossible
(c) Possible

E5 (a), (b), (d) and (f) are impossible.

E6 The pupil's own drawing

11 Re-arranging formulas(2)

This is a continuation of chapter 9.

A1 $g = f - h$

A2 $t = \dfrac{s - r}{x}$

A3 (a) $y = x - k$ (b) $y = \dfrac{a - b}{k}$
(c) $k = \dfrac{a - b}{y}$ (d) $x = \dfrac{ab - y}{p}$

A4 $m = a(c - d)$

A5 (a) $b = c(a - s)$ (b) $x = n(3 - f)$
(c) $z = 4(b - k)$ (d) $r = s(ap - q)$

A6 (a) $w = \dfrac{t + b}{a}$ (b) $m = uf$

(c) $s = \dfrac{p - r}{q}$ (d) $u = \dfrac{z}{w}$

(e) $m = \dfrac{y - c}{x}$ (f) $r = \dfrac{d - f}{p}$

(g) $s = n(k - m)$ (h) $x = s(a - w)$

B1 $y = \dfrac{v}{px}$

B2 $f = \dfrac{d}{hs}$

B3 $s = \dfrac{z}{mt}$

B4 (a) $p = \dfrac{t}{rs}$ (b) $g = \dfrac{m}{fkw}$

(c) $r = \dfrac{s}{3ty}$ (d) $b = \dfrac{k}{a^2 c}$

B5 $h = \dfrac{2A}{b}$

B6 (a) $T = \dfrac{100I}{PR}$ (b) $P = \dfrac{100I}{RT}$

(c) $R = \dfrac{100I}{PT}$ (d) $l = \dfrac{YAe}{F}$

(e) $F = \dfrac{YAe}{l}$ (f) $a = \dfrac{ktuv}{bs}$

(g) $m = \dfrac{fpqr}{l}$ (h) $h = \dfrac{3V}{\pi r^2}$

C1 (a) 10 (b) 1·2

C2 (a) $s = 10·125$ (b) $z = \dfrac{saf}{vy}$

(c) $z = 0·6$

C3 $h = 14·8$ (to 1 d.p.) so height is 14·8 cm

D1 (a) $x = 1·2$ (b) $y = 1·7$ (c) $p = 7$

D2 (a) $n = \dfrac{m}{lz}$ (b) $x = \dfrac{be}{ca}$

(c) $n = \dfrac{A}{5f}$ (d) $r = \dfrac{ut}{a}$

D3 (a) $w = \dfrac{s-t}{uv}$ (b) $w = \dfrac{t-s}{uv}$

(c) $R = \dfrac{PT}{F}$ (d) $l = a(m+b)$

(e) $b = \dfrac{l}{a} - m$ (f) $b = \dfrac{5p}{q^3}$

(g) $x = n(y - ab)$ (h) $x = n(ab - y)$

(i) $u = \dfrac{bv(t-s)}{a}$ or $\dfrac{bv}{a}(t-s)$

*⋆***D4** (a) $h = \dfrac{f - ag}{a}$ (b) $q = \dfrac{m + pr}{p}$

(c) $c = \dfrac{ab - d}{a}$ (d) $t = \dfrac{fgs - h}{fg}$

(e) $b = \dfrac{a - dc}{d}$ (f) $x = \dfrac{a - by}{y}$

(g) $x = \dfrac{by - a}{y}$ (h) $r = \dfrac{b}{a - s}$

Review 2

---

6 Trigonometry

6.1 a 3·6 cm, b 3·9 cm, c 4·2 cm
(all to 1 d.p.)

6·2 a 5·0 cm, b 8·2 cm, c 11·1 cm
(all to 1 d.p.)

6.3 $a = 65·4°$, $b = 24·8°$, $c = 34·1°$
$d = 39·6°$, $e = 29·1°$, $f = 60·9°$

6.4 64°

6.5 (a) 7·0 km
(b) 4·7 km
(c) 1·5 km
(d) 8·7 km
(e) 9·9 km

6.6

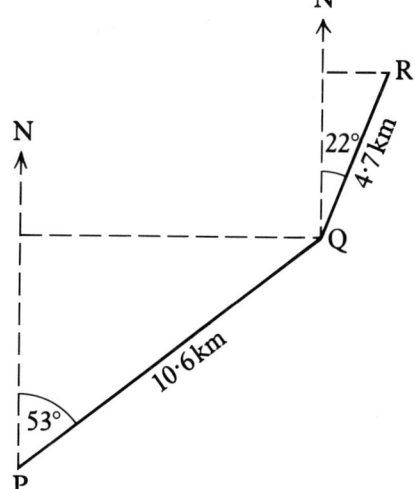

(a) 10·7 km north (b) 10·2 km east
(10·8 km and 10·3 km if you round
early)

6.7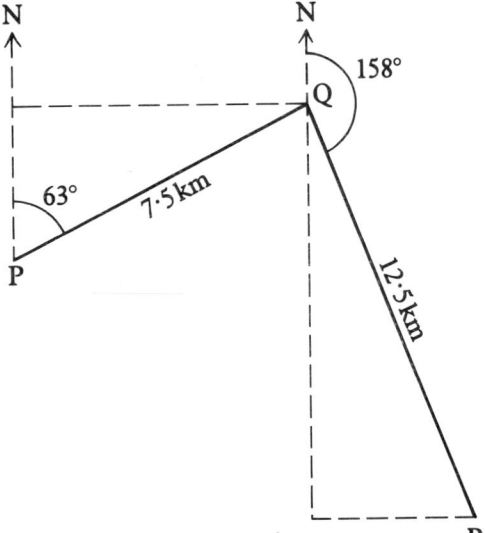

(a) $^-$8·2 km north   (b) 11·4 km east

6·8 45·6°, 45·6° and 88·8° (to 3 s.f.)

6·9 90·0° (to 3 s.f.)

8 Distributions

8.1

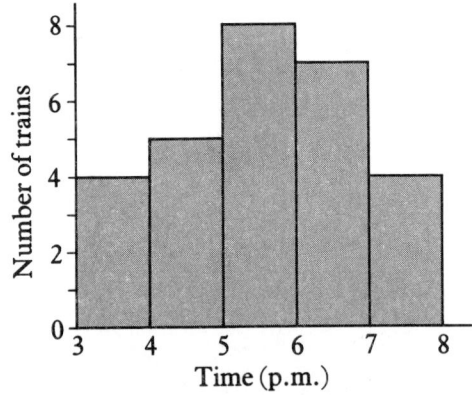

Time (p.m.)

Train on the hour included in the next interval

8.2 (a)

Journey time in minutes	Frequency
37	1
38	4
39	3
40	1
42	1
44	1
46	1
47	4
48	2

(b) 42·3 minutes (to 1 d.p.)

(c) No, the mean value does not represent a typical journey. There are basically two types of trains – fast ones and slow ones.

(d) The range is 11 minutes.

8.3 4·3 letters per word (to 2 s.f.)

8.4 6·2 peas in a pod.

8.5 (a) 3·0 kg (to 2 s.f.) (b) 19%

9 Re-arranging formulas (1)

9.1 (a) $u = {}^-58·7$ (b) $a = 1·3$

9.2 (a) $q = p + r - s$ (b) $s = p - q + r$

9.3 (a) $t = \dfrac{y + b}{a}$ (b) $k = n(h - m)$

(c) $u = 3(s + a)$ (d) $x = a(z - y)$

(e) $u = \dfrac{d - p}{v}$ (f) $x = s(w - k)$

10 Points, lines and planes

10.1 A and C are above,   B and D are below.

10.2 D must be above the plane to go to E in a straight line.
D must be below the plane to go to C in a straight line.

10.3 (a)

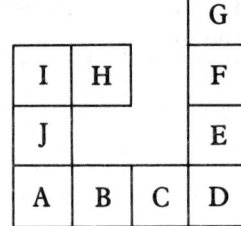

(b) H and G are not next to each other.

11 Re-arranging formulas (2)

11.1 (a) $q = \dfrac{p + s}{r}$ (b) $s = qr - p$

(c) $b = \dfrac{a - d}{c}$ (d) $w = \dfrac{u}{v} - t$

(e) $u = v(t + w)$ (f) $g = a(f - h)$

(g) $x = \dfrac{z}{3uy}$ (h) $p = \dfrac{rs}{2q}$

(i) $d = \dfrac{5a^2b}{ct}$

11.2 (a) 34·4 stones (b) $l = \dfrac{21w}{5g^2}$

(c) $l = 4·2$ feet

M Miscellaneous

M1 7 shapes excluding rotations and reflections

12 Proportionality

Proportionality was introduced in chapter 1. This present chapter looks at further aspects of this type of relationship, including the use of algebra.

Proportionality is a very important concept in science, especially physics. In experimental work one would not expect to get exact proportionality, hence the technique described in section E.

A1 (a)

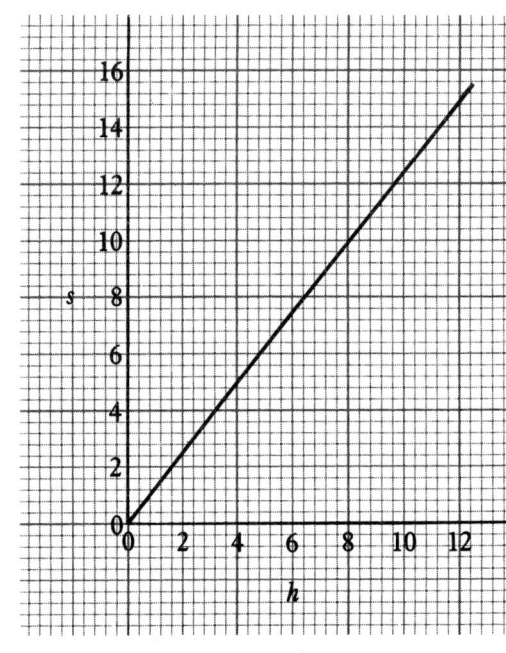

(b) (i) 6·0m (ii) 14·7m
(c) 8·7m

A2 (a) 14m (b) 35m

A3 486cm

A4 34·8m

A5 16·6m

A6 278cm^3 (to 3 s.f.)

A7 12·8cm (to 3 s.f.)

A8 586 (to 3 s.f.)

A9 42·3 (to 3 s.f.)

A10 (a) 6·9 (b) 4·8
(c) 6·3 (d) 20·7
(e) 8·0 (f) 6·5

B1 (a)

s	0	1	2	3	4	5	6
A	0	1	4	9	16	25	36

(b)

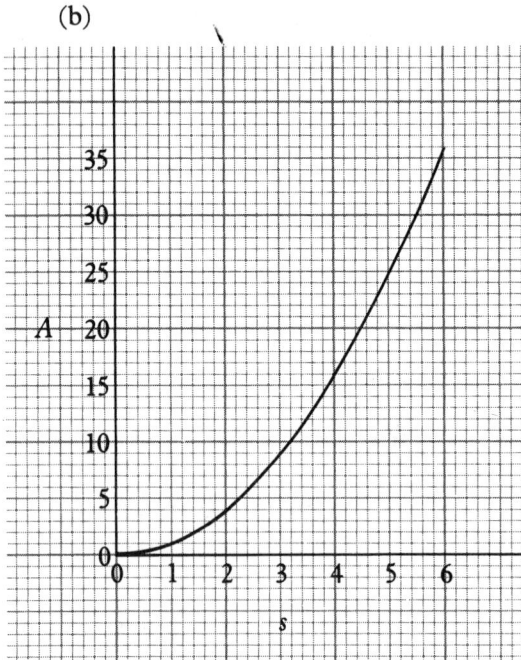

(c) A is not proportional to s.
(d) When s is doubled, A increases by a factor of 4.

B2 (a) No (b) No (c) No
(d) See part (f)
(e) Cost is not proportional to the quantity. The graph is not a straight line through the origin.
(f)

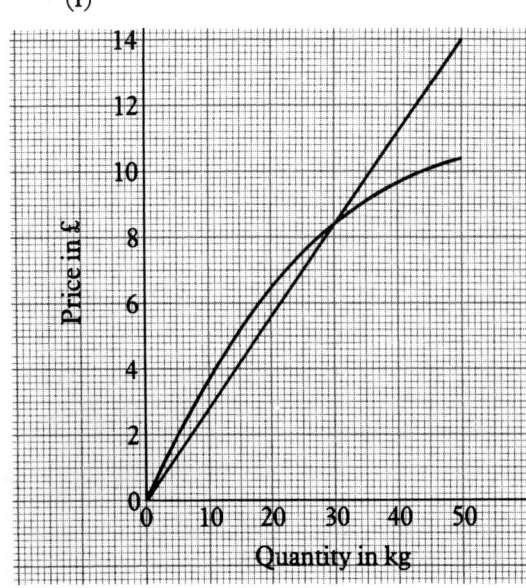

(g)

Quantity in kg	5	10	15	30	40	50
Price in £	1·40	2·80	4·20	8·40	11·20	14·00

B3

$$p \ 7\cdot61 \ \xrightarrow{\times 10} \ 76\cdot10$$

$$q \ 27\cdot72 \ \xrightarrow[\times \ ?]{} \ 233\cdot19$$

? is not equal to 10.

C1 (a) 30 m (b) 1·5
(c) 45 m (d) 1·5
(e) 1·5

C2 (a) 16 (b) 0·8
(c) 0·8 (d) 0·8

C3 A 0·5 B 0·6,
C 0·7, D 0·8

C4 A 0·9, B 0·73,
C 0·6, D 0·57

C5 (a)

p	550	635	725	880	1210
q	770	886	1015	1408	1815
$\dfrac{q}{p}$	1·400	1·395	1·400	1·600	1·500

(b) q is not proportional to p.

C6 Yes, V is proportional to I.
The ratio $\dfrac{V}{I}$ is constant.

D1 (a) (i) 4 (ii) $q = 4p$ (iii) 800
(b) (i) 2 (ii) $q = 2p$ (iii) 400
(c) (i) 1·5 (ii) $q = 1\cdot5p$ (iii) 300
(d) (i) 1·2 (ii) $q = 1\cdot2p$ (iii) 240
(e) (i) 1 (ii) $q = p$ (iii) 200
(f) (i) 0·75 (ii) $q = 0\cdot75p$ (iii) 150
(g) (i) 0·6 (ii) $q = 0\cdot6p$ (iii) 120
(h) (i) 0·3 (ii) $q = 0\cdot3p$ (iii) 60
(i) (i) 0·1 (ii) $q = 0\cdot1p$ (iii) 20

D2 (a)

s	20	25	35	45
l	32	40	56	72

(b)

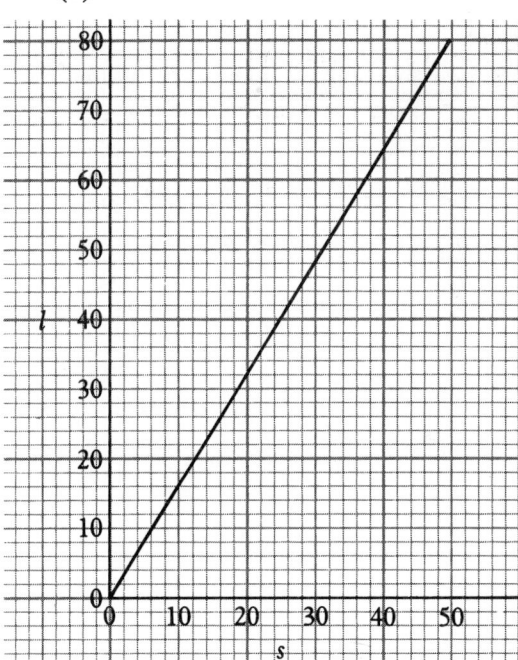

(c) Due to small measurement errors the values of s and l are not quite in proportion.

(d) 1·6 (e) $l = 1·6s$

D3 (a)

	A	B	C	D
s	30	42	50	70
l	36	54	60	84

(b)

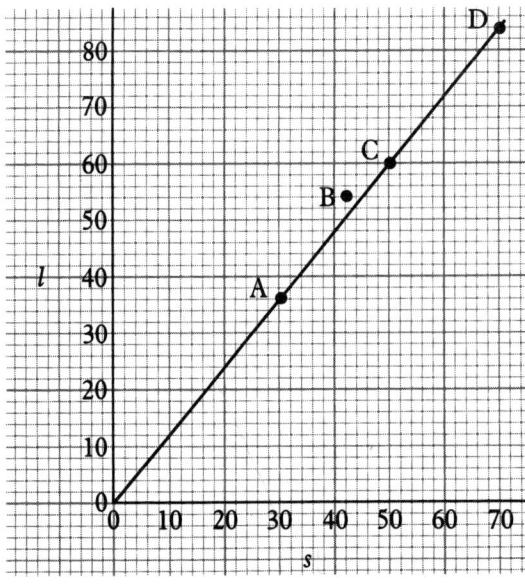

Rectangle B is the 'odd one out'.

(c) $l = 1·2s$ (d) 1·286 (to 4 s.f.)

D4 (a)

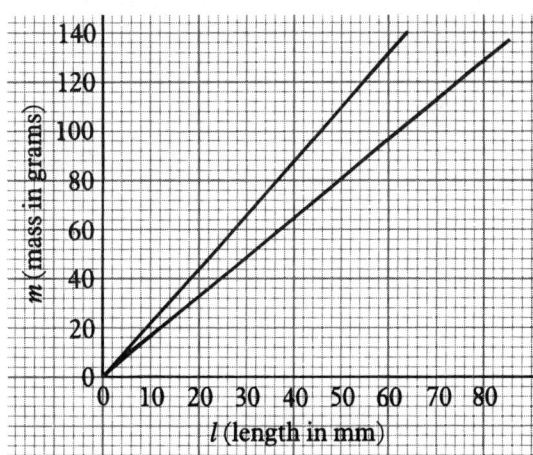

(b) $m = 1·6l$ and $m = 2·2l$. The second represents the thicker wire.

E1 (a) 1·6, 1·8, 1·67, 1·75, 1·74, 1·67
(b) Three are greater than 1·7, and three are less than 1·7.

E2 (a) and (b)

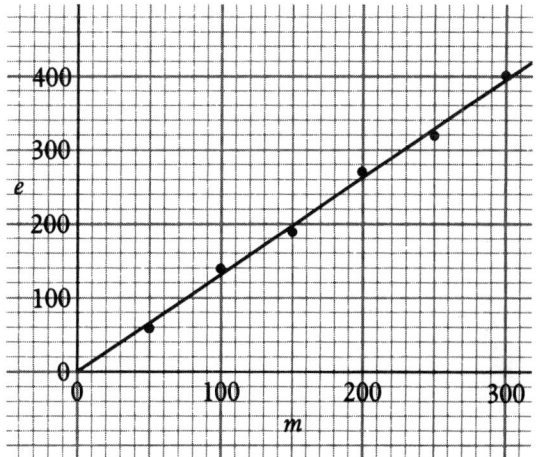

(c) Gradient 1·3 (to 2 s.f.), $e = 1·3m$

E3 (a)

m	10	20	30	40	50
e	14	25	42	52	63

(b)

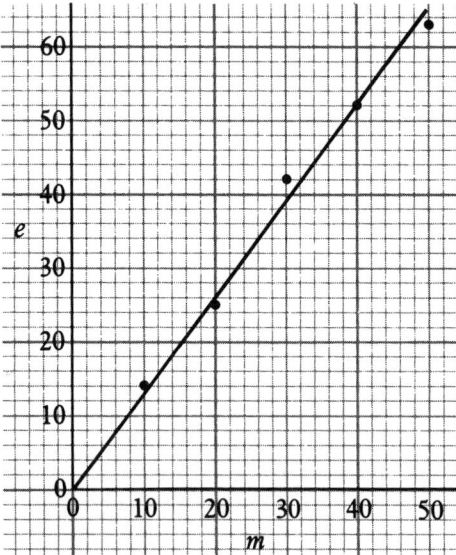

(c) Gradient 1·3 (to 2 s.f.), $e = 1·3m$

E4   (a)

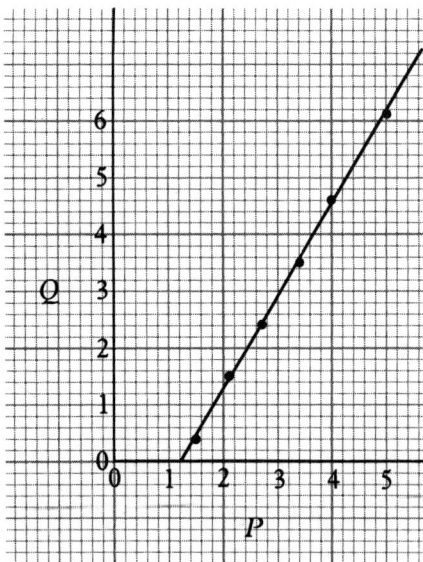

(b) Q is not proportional to P.

★E5   (a)

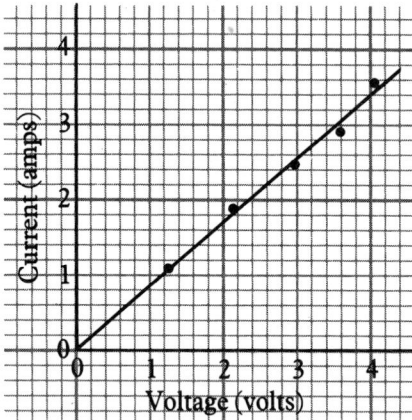

(b) Yes
(c) 1·2 (to 1 d.p.)

13 Area

This chapter covers the area of the parallelogram, triangle, trapezium and circle. The more usual explanation of the rule for the area of a parallelogram, shown in the sequence of diagrams below, breaks down when the 'top' of the parallelogram completely overhangs the base.

  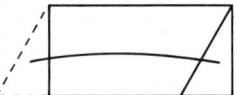

A1 $60\,\text{cm}^2$

A2 (a) $12\cdot7\,\text{cm}^2$
(b) Rounding and measurement errors may give you a slightly different answer here.

A3 (a) 21 sq units   (b) 14 sq units
(c) 30 sq units

A4 $13\cdot5\,\text{cm}^2$

A5 $66\,\text{cm}^2$

A6 d is 24 cm

B1 (a), (b), (c) Answers will vary from about $10\cdot5$ to $11\cdot0$.

B2 (a) $12\frac{1}{2}$ sq units   (b) 7 sq units
(c) $10\frac{1}{2}$ sq units

B3 d is 9·6 cm.

C1 B and D are trapeziums.

C2 (a) $5\,\text{cm}^2$ (b) $3\,\text{cm}^2$ (c) $8\,\text{cm}^2$

C3 (a) $39\,\text{cm}^2$ (b) $35\,\text{cm}^2$ (c) $33\,\text{cm}^2$

C4 40 sq units

C5 (a) $306\,\text{cm}^2$ (b) $12\cdot825\,\text{cm}^2$

C6 $22\cdot6\,\text{m}^2$

C7 (a) A $76\cdot075\,\text{m}^2$,   B $30\,\text{m}^2$,
C $77\cdot88\,\text{m}^2$,   D $21\cdot6\,\text{m}^2$
(b) $205\cdot555\,\text{m}^2$ (c) £1336 (to 4 s.f.)

C8 $218\cdot2\,\text{m}^2$

C9 $744\,\text{m}^2$

D1 (a) (i) 45 cm   (ii) $155\cdot25\,\text{cm}^2$
(b) (i) 36·4 cm   (ii) $98\cdot28\,\text{cm}^2$
(c) (i) 37·2 cm   (ii) $107\cdot88\,\text{cm}^2$

D2 (a) $78\cdot5\,\text{m}^2$ (b) $141\cdot0\,\text{m}^2$ (c) $572\cdot6\,\text{m}^2$
(d) $3097\cdot5\,\text{m}^2$ (e) $125\,663\cdot7\,\text{m}^2$

D3 (a) 30 m   (b) $72\,\text{m}^2$

D4 (a) 42·7 cm   (b) $145\,\text{cm}^2$

D5 (a) $21\cdot5\,\text{m}^2$ (b) $101\,\text{m}^2$
(c) $79\cdot3\,\text{m}^2$ (all to 3 s.f.)

E1 $r = 1\cdot99$

E2 $r = 2\cdot47$

E3 2·8 cm

E4 8·0 cm

E5 (a) Pupil's guess   (b) 1·46 m

14 Linear equations and inequalities

This chapter is mainly concerned with equations of the form $ax + by = c$, and their graphs, and inequalities of the forms $ax + by < c$, etc., and their representation as regions. The chapter includes the graphical solution of simultaneous linear equations. The algebraic method of solution is dealt with in a later book.

A1 A, D and E are portrait.
C and G are square.
B, F and H are landscape.

A2 (a)

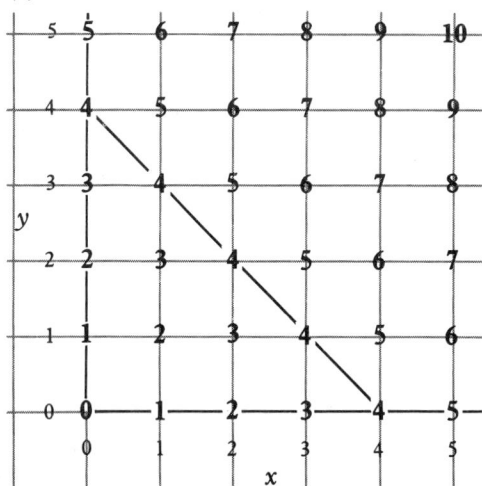

(b) $x + y < 4$ (c) $x + y > 4$

A3

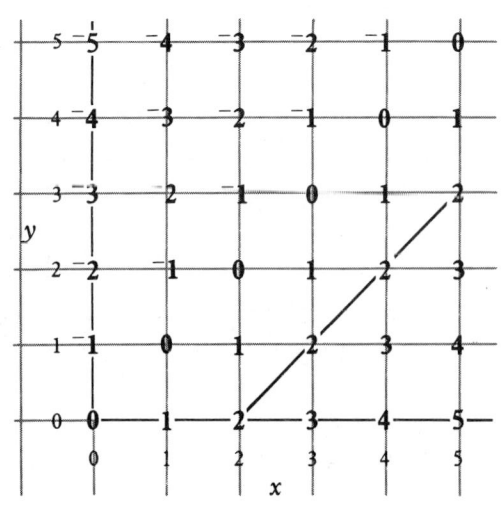

(a) $x - y > 2$ (b) $x - y < 2$

A4

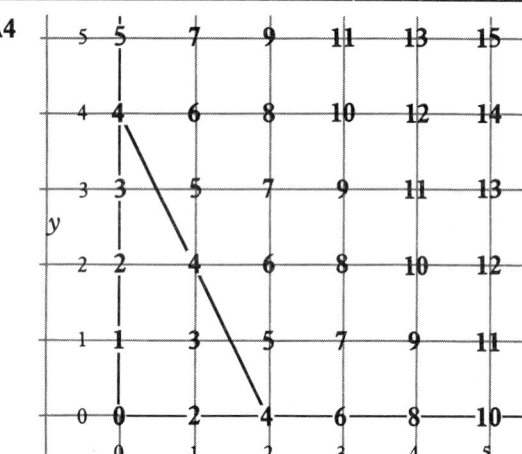

(a) $2x + y < 4$ (b) $2x + y > 4$

A5 (a) $y > 2x$ (b) $y < 2x$

A6 (a) $y = \frac{1}{3}x$ (b) $y < \frac{1}{3}x$ (c) $y > \frac{1}{3}x$

B1

B2 (a), (c) and (d) qualify.

B3 4 large bags

30

B4   5 small bags

B5

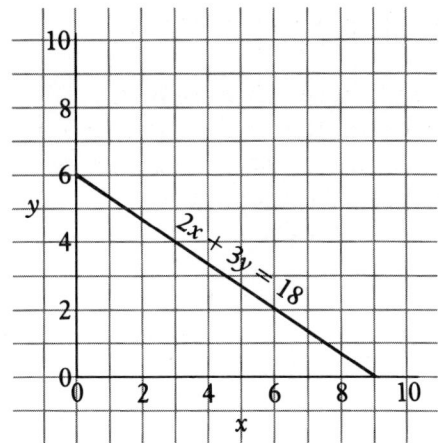

B6   (a), (b), (c) and (e)

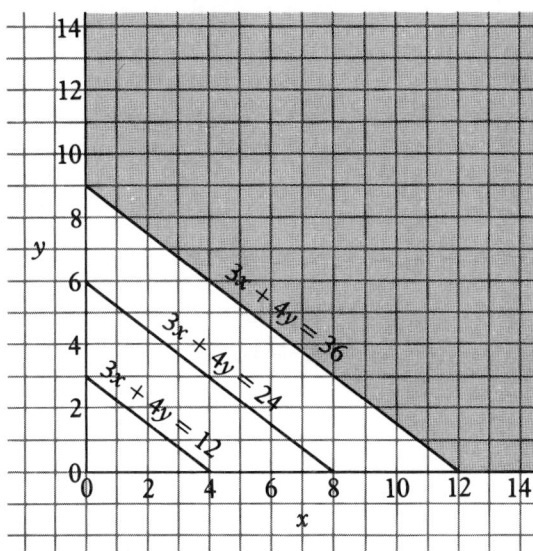

(d)   The three lines are parallel.

B7

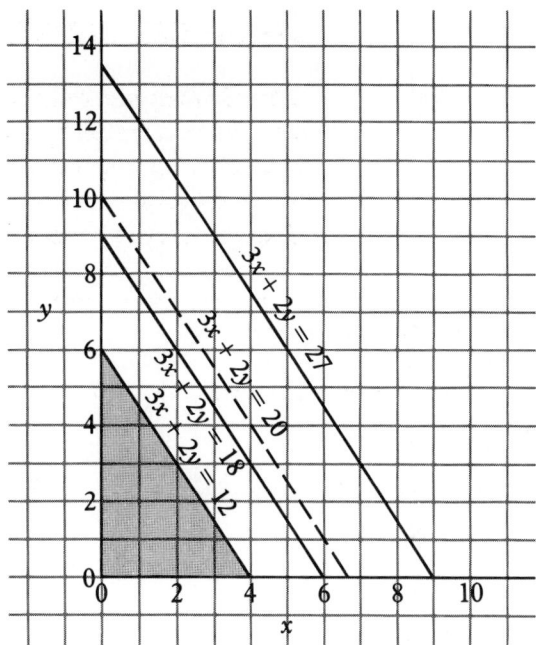

C1   (a)   Standard price
    (b)   Discount
    (c)   Delivery charge

C2   (a) $4x + 3y$
    (b) $4x + 3y = 24$
    (c)

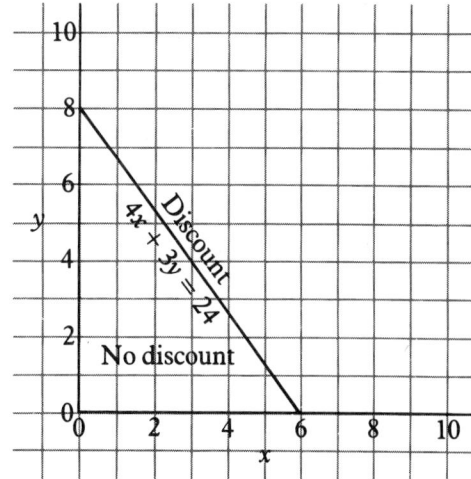

C3   (a) $8x + 6y$
    (b) $8x + 6y = 72$

(c)

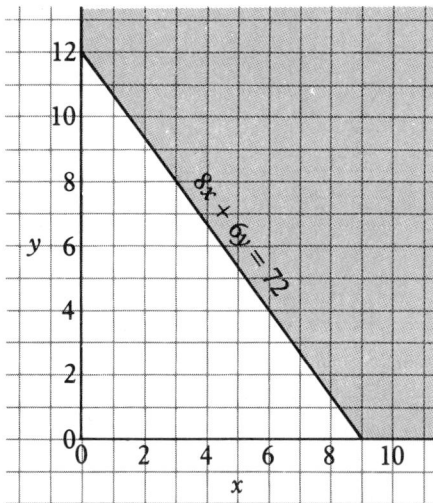

C4 (a) $5x + 3y$ (b) $5x + 3y = 15$
(c) $5x + 3y = 30$
(d)

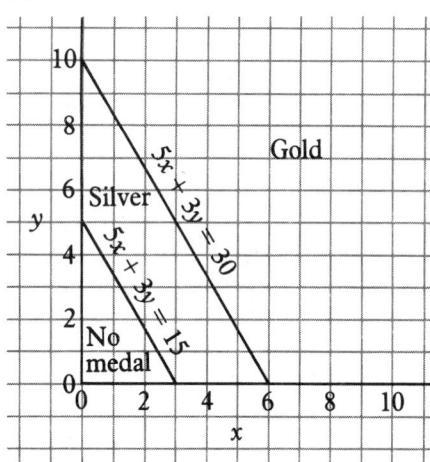

D1 $(2 \times 6) + (3 \times 4) = 24, \quad 6 + 8 = 14$

D2 (a) and (b)

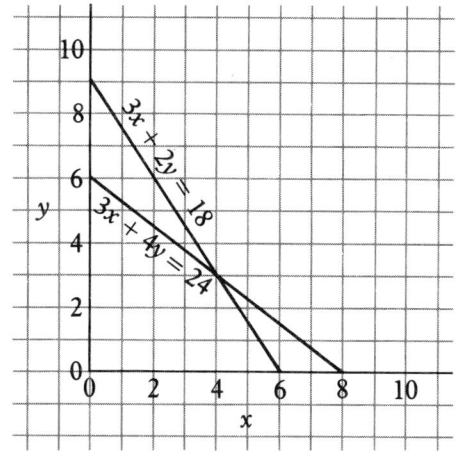

(c) $x = 4$ and $y = 3$
(d) $(3 \times 4) + (2 \times 3) = 18,$
$(3 \times 4) + (4 \times 3) = 24$

D3 (a)

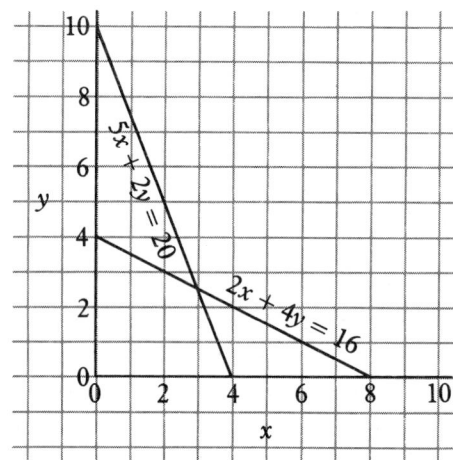

(b) $x = 3$ and $y = 2\frac{1}{2}$
(c) $(2 \times 3) + (4 \times 2\frac{1}{2}) = 16$
$(5 \times 3) + (2 \times 2\frac{1}{2}) = 20$

D4 (a) $x = 0, \quad y = 6$
(b) $x = 6, \quad y = 3$
(c) $x = 4, \quad y = 2$

D5 (a) and (b)

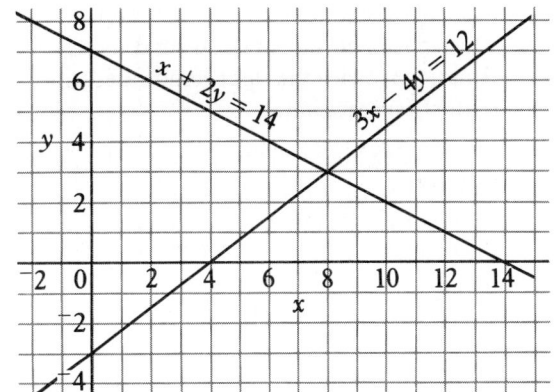

(b)

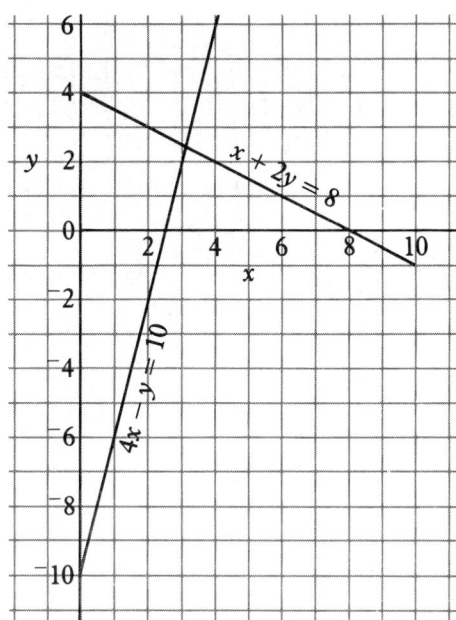

$x = 3 \cdot 1, \quad y = 2 \cdot 4 \text{ (to 1 d.p.)}$

(c) $x = 8, \quad y = 3$

(d) $8 + (2 \times 3) = 14,$
$(3 \times 8) - (4 \times 3) = 12$

D6 (a)

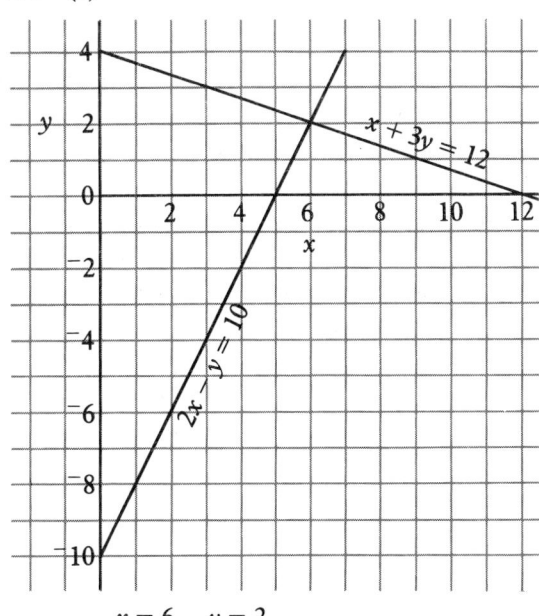

$x = 6, \quad y = 2$

(c)

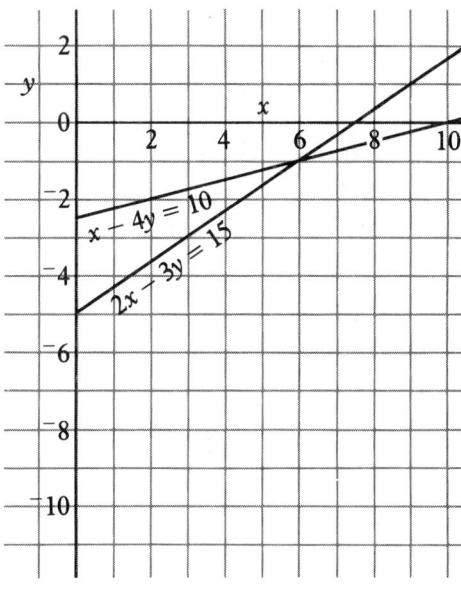

$x = 6, \quad y = {}^-1$

(d)

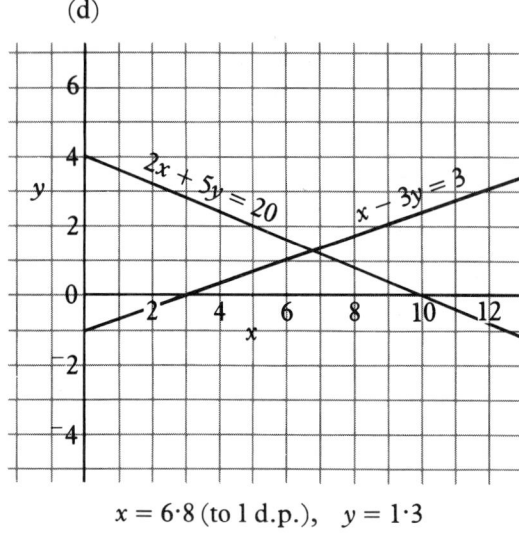

$x = 6\cdot8$ (to 1 d.p.), $y = 1\cdot3$

D7 The two lines are parallel.
The pair of equations has no solution.
The right-hand side of the second equation is twice the right-hand side of the first equation.

D8 If you halve both sides of the second equation you get the first equation. The first equation is equivalent to the second.

15 Investigations(2)

See the note on chapter 7 for the purpose of investigations.

1

Height in cm	Number of ways
1	2
2	4
3	8
4	16
.	.
.	.
.	.

To each of the arrangements of the previous height you can add either a red or a white cube.

2 There are 64 ways of arranging the six cubes.
For a box containing n cubes there will be 2^n arrangements.

★3 16 different blocks can be made.

16 Periodic graphs

This chapter is concerned with examples of periodic and approximately periodic phenomena. The most important mathematical instances of a periodic graph – the sine and cosine graphs – are dealt with in a later book.

A1 35 minutes

A2 (a) 45 min (b) 60 min

B1

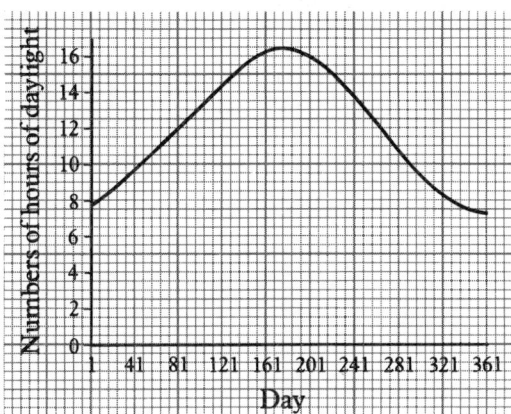

B2 1 year

B3 (a) 1 year
  (b) The variation in length of day
      between summer and winter is less
      than in London. Long hours of day-
      light in London correspond to short
      hours of daylight in Melbourne.
  (c) It is explained on page 136 of
      the pupil's book.

B4 C, D, B, E, A, F

B5 On the equator

B6

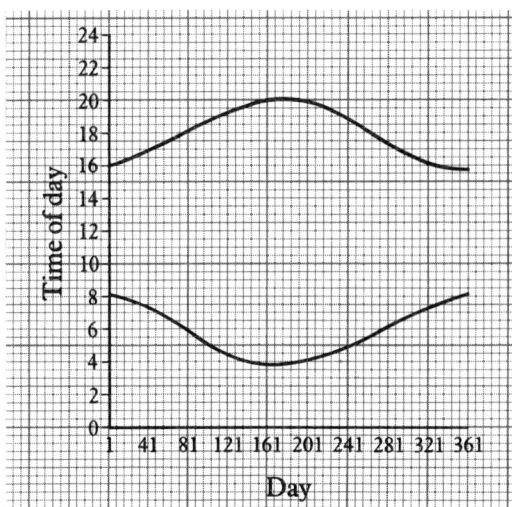

B7 (a) In the southern hemisphere
  (b) On the equator
  (c) Close to the north pole

C1 (a) The height of high tide and low tide is
      different on different days.
  (b) 29 low tides
  (c) 12·4 hours

C2 (a) The 9th day
  (b) 2nd or 3rd day

C3 (a) 15th March
  (b) 8th and 21st–22nd March
  (c) About 45 feet
  (d) About 20 feet

C4 The graph is steeper for the incoming tide
     than for the outgoing tide.

C5 (a) About 3 hours
  (b) About 7 hours

C6 (a) 24 hours
  (b) The depth is greater than 1·5 m for
      about 22 hours.

C7 The ship will have to wait about $3\frac{1}{4}$ hours
     for there to be a depth of 4 metres, but the
     captain will probably wait longer to give a
     safety margin of depth.

C8 (a) About 7:30 a.m.
  (b) About 6 hours
  (c) 2·2 m, on Monday.

C9 Wednesday is probably best because the
     road is clear for longer in the afternoon.

C10 About 6:00 a.m.

C11 About 6 hours

C12 About 12 noon

*D1 (a) Group flashing 15 seconds
  (b) Occulting, 12 seconds
  (c) Interrupted quick flashing, 9 seconds
  (d) Quick flashing, $\frac{1}{2}$ second
  (e) Flashing, 6 seconds
  (f) Fixed and flashing, 5 seconds
  (g) Group occulting 20 seconds
  (h) Isophase, 8 seconds
  (i) Fixed, no period

★D2

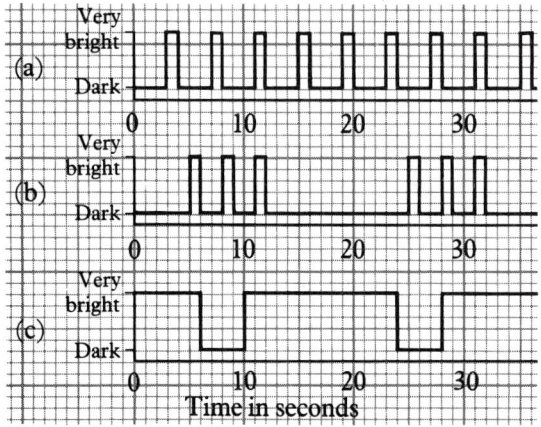

17 Probability

Probability was first introduced in *Book Y1*, through the idea of relative frequency. (See the note on chapter 12 of *Book Y1*.) This chapter introduces the idea of equally likely outcomes.

A1 $\frac{3}{50}$ or 0·06

A2 (a) $\frac{1}{40}$, 0·025 (b) $\frac{7}{40}$, 0·175
(c) $\frac{7}{400}$, 0·0175 (d) $\frac{3}{1000}$, 0·003
(e) $\frac{1}{65}$, 0·0154 (to 3 s.f.)
(f) $\frac{9}{65}$, 0·138 (to 3 s.f.)
(g) $\frac{1}{100}$, 0·01 (h) $\frac{1}{100}$, 0·01

A3 (a) $\frac{1}{20}$, 0·05 (b) $\frac{19}{20}$, 0·95

A4 (a) $\frac{1}{6}$ (b) $\frac{2}{3}$ (c) $\frac{1}{3}$

A5 (a) $\frac{1}{8}$ (b) $\frac{3}{8}$ (c) $\frac{1}{2}$ (d) $\frac{1}{8}$

A6 $\frac{15}{88} = \frac{3}{16}$

B1 $\frac{3}{4}$

B2 $\frac{3}{8}$

B3 (a) $\frac{1}{4}$ (b) $\frac{3}{8}$ (c) $\frac{1}{2}$

B4 (a) $\frac{3}{8}$ (b) $\frac{3}{8}$

B5 (a) $\frac{9}{100}$ (b) $\frac{9}{10}$ (c) $\frac{1}{100}$ (d) 0

B6 The girl is correct. Each ticket has the same chance of winning. The winning number is more likely to be a two-figure number because there are more of them but the boy has only one such ticket so the probability of a winning two-figure number being his is small.

C1 (a) $\frac{1}{4}$ (b) $\frac{1}{4}$

C2 (a)

A	B	C
H	H	H
H	H	T
H	T	H
H	T	T
T	H	H
T	H	T
T	T	H
T	T	T

(b) 3 (c) $\frac{3}{8}$ (d) $\frac{3}{8}$ (e) (i) $\frac{1}{8}$ (ii) $\frac{1}{8}$

C3 (a)

H	H	H	H		T	H	H	H
H	H	H	T		T	H	H	T
H	H	T	H		T	H	T	H
H	H	T	T		T	H	T	T
H	T	H	H		T	T	H	H
H	T	H	T		T	T	H	T
H	T	T	H		T	T	T	H
H	T	T	T		T	T	T	T

(b) (i) $\frac{1}{16}$ (ii) $\frac{1}{4}$ (iii) $\frac{3}{8}$ (iv) $\frac{1}{4}$ (v) $\frac{1}{16}$

C4 (a) 32 (b) 1 (c) $\frac{1}{32}$ (d) $\frac{1}{128}$

D1 (a)

A B	A B	A B	A B	A B	A B
1 1	2 1	3 1	4 1	5 1	6 1
1 2	2 2	3 2	4 2	5 2	6 2
1 3	2 3	3 3	4 3	5 3	6 3
1 4	2 4	3 4	4 4	5 4	6 4
1 5	2 5	3 5	4 5	5 5	6 5
1 6	2 6	3 6	4 6	5 6	6 6

(b) 36 (c) 4 (d) $\frac{4}{36} = \frac{1}{9}$ (e) 3
(f) $\frac{1}{12}$ (g) Score of 7, $\frac{1}{6}$

D2 (a)

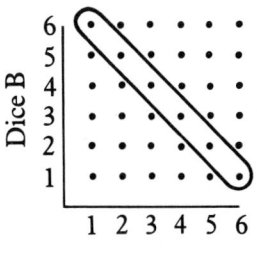

Dice A

(b) $\frac{1}{6}$

D3 (a) $(5, 4)$, $(6, 3)$ and $(3, 6)$
(b) $\frac{4}{36} = \frac{1}{9}$

D4

Total score with 2 dice	2	3	4	5	6	7	8	9	10	11	12
Probability	$\frac{1}{36}$	$\frac{2}{36}$	$\frac{3}{36}$	$\frac{4}{36}$	$\frac{5}{36}$	$\frac{6}{36}$	$\frac{5}{36}$	$\frac{4}{36}$	$\frac{3}{36}$	$\frac{2}{36}$	$\frac{1}{36}$

D5 (a) 6 (b) $\frac{1}{6}$
(c) $\frac{3}{36} + \frac{2}{36} + \frac{1}{36} = \frac{6}{36} = \frac{1}{6}$

D6 (a)

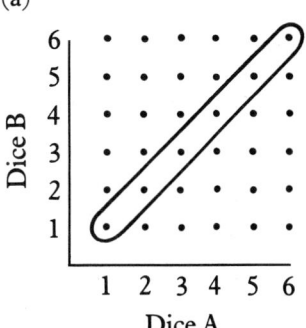

Dice A

(b) $\frac{1}{6}$

D7 (a)

```
6 | .5 .4 .3 .2 .1 .0
5 | .4 .3 .2 .1 .0 .1
4 | .3 .2 .1 .0 .1 .2
3 | .2 .1 .0 .1 .2 .3
2 | .1 .0 .1 .2 .3 .4
1 | .0 .1 .2 .3 .4 .5
  +------------------
    1  2  3  4  5  6
```
Dice B (vertical axis) Dice A (horizontal axis)

(b) 8
(c) $\frac{8}{36} = \frac{2}{9}$
(d)

Difference between scores on two dice	0	1	2	3	4	5
Probability	$\frac{6}{36}$	$\frac{10}{36}$	$\frac{8}{36}$	$\frac{6}{36}$	$\frac{4}{36}$	$\frac{2}{36}$

E1 (a)

I	II	III
I	III	II
II	I	III
II	III	I
III	I	II
III	II	I

(b) $\frac{1}{6}$
(c) $\frac{1}{3}$

E2 (a) $\frac{6}{24} = \frac{1}{4}$
(b) $\frac{6}{24} = \frac{1}{4}$
(c) $\frac{1}{2}$ (d) $\frac{1}{24}$
(e) $\frac{1}{12}$

E3 (a) $\frac{1}{2}$ (b) $\frac{1}{2}$ (c) $\frac{1}{3}$

Review 3

12 Proportionality

12.1 (a) Graph A
(b) The cooking time is not proportional to the weight.

12.2

p	3·8	5·3	⑨·⑥
q	10·1	⑭·①	25·6

12.3 23·5 minutes (to 3 s.f.)

12.4 $v = 1·4u$

12.5 (a), (b) and (c)

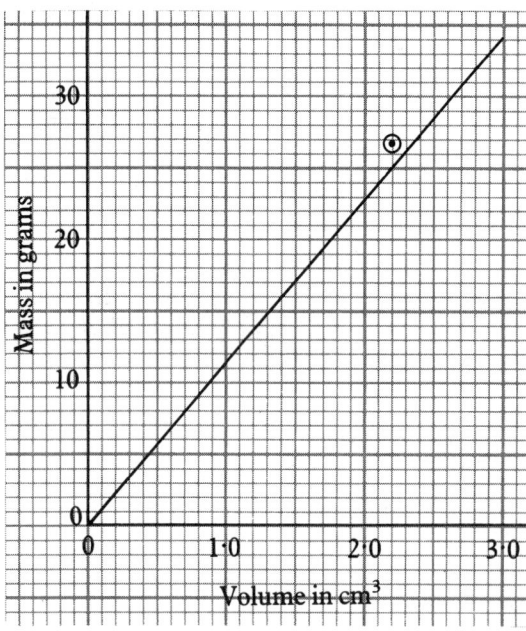

(d) 11·3
(e) The density of lead is 11·3 grams per cm^3.

13 Area

13.1 (a) 26·52 cm^2 (b) 9·9 cm^2
(c) 50·88 cm^2 (d) 36·54 cm^2

13.2 14·96 cm^2

13.3 940 m^2

13.4 (a) 105·7 cm^2 (b) 26·4 cm^2
(c) 2·5 cm

13.5 (a) 7·96 m (to 3 s.f.)
(b) 199 m^2 (to 3 s.f.)

13.6 (a) 40 m (b) 35·45 m (to 2 d.p.)
(c) 78·5 m^2 (to 1 d.p.)

14 Linear equations and inequalities

14.1 (a) and (b)

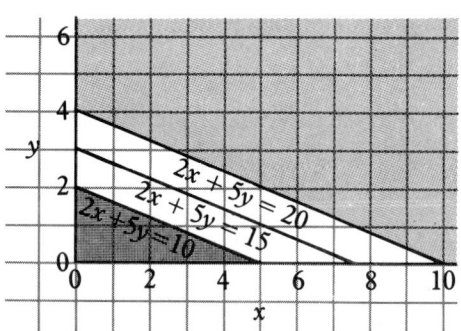

14.2 (a) $6x + 3y$ (b) $6x + 3y = 15$
(c) $6x + 3y = 30$
(d)

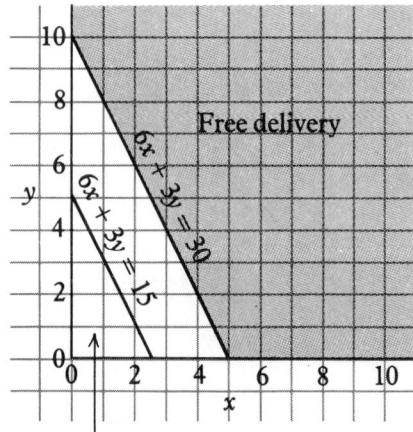

Not accepted

38

14.3

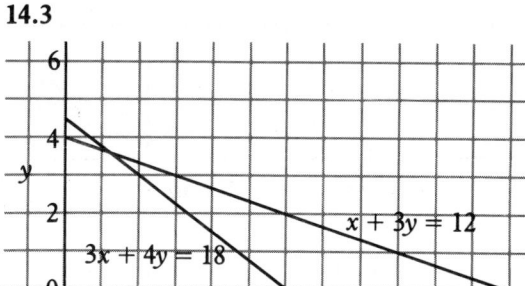

$$x = 1 \cdot 2, \quad y = 3 \cdot 6$$

14.4 (a)

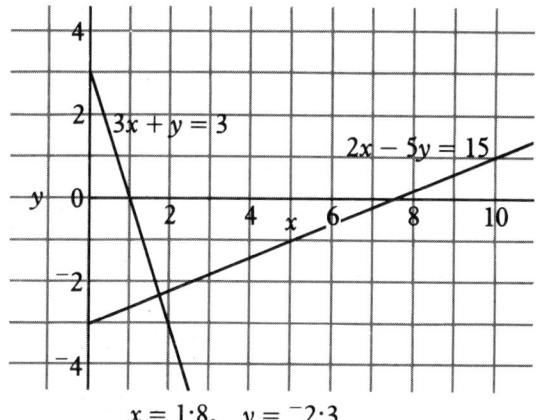

$$x = 1 \cdot 8, \quad y = {}^-2 \cdot 3$$

(b)

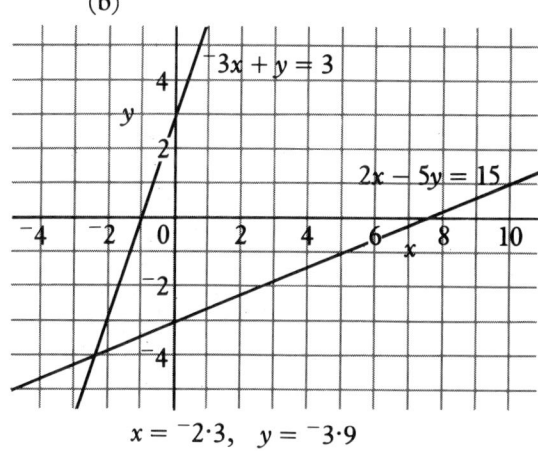

$$x = {}^-2 \cdot 3, \quad y = {}^-3 \cdot 9$$

16 Periodic graphs

16.1

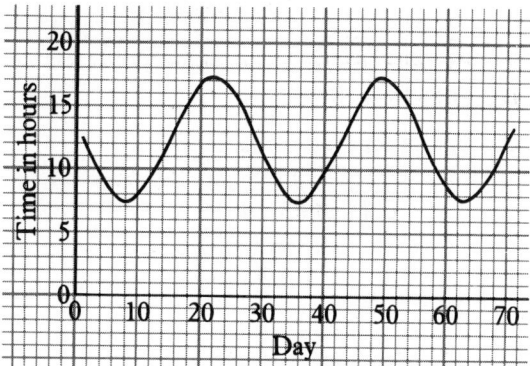

The period is about 28 days.

16.2

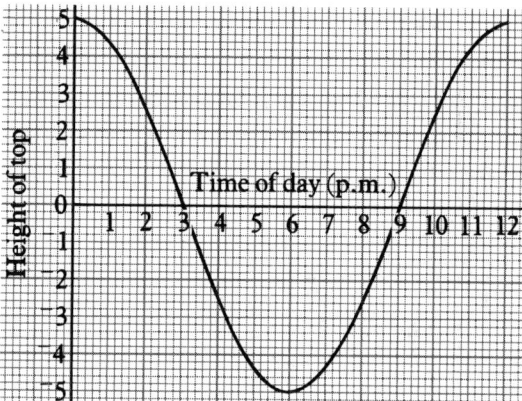

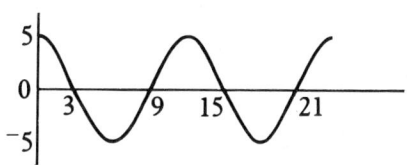

17 Probability

17.1 (a) $\frac{1}{6}$ (b) $\frac{1}{6}$ (c) $\frac{1}{12}$ (d) $\frac{1}{4}$
(e) $\frac{1}{4}$

17.2 (a) and (b)

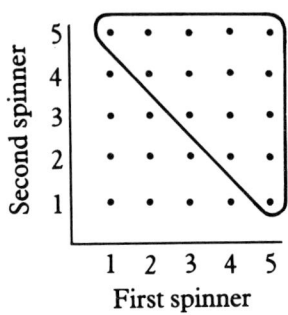

(c) $\frac{3}{5} = 0.6$

M Miscellaneous

M1 (a) B (b) C (c) A and C

(d)

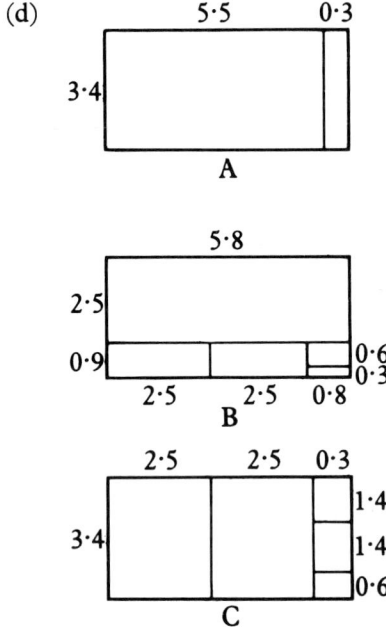

Published by the Press Syndicate of the University of Cambridge
The Pitt Building, Trumpington Street, Cambridge CB2 1RP
40 West 20th Street, New York, NY 10011–4211, USA
10 Stamford Road, Oakleigh, Melbourne 3166, Australia

© Cambridge University Press 1986

First published 1986
Sixth printing 1993

Typesetting and diagrams by Marlborough Design
Printed in Great Britain at the University Press, Cambridge

ISBN 0 521 31672 3

ED